RICK PETERS

ELECTRICAL BASICS

Main Street
A division of Sterling Publishing Co., Inc.
New York

Acknowledgements

Butterick Media Production Staff

Photography: Christopher Vendetta
Cover Photo: Brian Kraus
Illustrations: Mario Camacho
Design: Triad Design Group, Ltd.
Photography Editor: Tony O'Malley
Coordinator: Danielle Tringali
Indexer: Nan Badgett
Copy Editor: Barbara Webb
Page Layout: David Joinnides
Associate Managing Editor: Stephanie Marracco
Project Director: Caroline Politi
President: Art Joinnides

Special thanks to Keith Kauffman, licsensed master electrician and electrical contractor, and the crew of Kauffman Electric for providing numerous props for photography and for reviewing the manuscript and gently pointing out technical corrections. Thanks to the production staff at Butterick Media for their continuing support. And finally, a heartfelt thanks to my constant inspiration: Cheryl, Lynne, Will and Beth. **R.P.**

10 9 8 7 6 5 4 3 2 1

Library of Congress Cataloging-in-Publication Data
Peters, Rick.
Electrical Basics/Rick Peters
p.cm.
ISBN 0-8069-3667-3
I. Electric wiring, Interior. I. Title.
TK3285.P48 2000
621.319'24--dc21 99-086645

ISBN 1-4027-1086-0

Published by Sterling Publishing Company, Inc.
387 Park Avenue South, New York, N.Y. 10016

Distributed in Canada by Sterling Publishing
c/o Canadian Manda Group, One Atlantic Avenue, Suite 105
Toronto, Ontario, Canada M6K 3E7
Distributed in Great Britain by Chrysalis Books
64 Brewery Road, London N7 9NT, England
Distributed in Australia by Capricorn Link (Australia) Pty. Ltd.
P.O. Box 704, Windsor, NSW 2756 Australia

Printed in China

Main Street
A division of Sterling Publishing Co., Inc.
New York

Contents

Introduction

Try this little experiment—wander through the rooms of your home and count how many items require electricity to function. I'd be surprised if your total is less than one hundred. Now try to remember that last time your home lost power. Inconvenient? Downright uncomfortable, perhaps? You bet.

It's quite staggering to realize how much we've come to depend on electrical devices: TVs, VCRs, computers, ranges, refrigerators, microwave ovens, hair dryers, toasters, blenders, garage door openers, coffee makers, air conditioners, lights, and radios, to name a few. Many homeowners also depend on electricity to heat and cool their homes and provide hot water.

Obviously you don't have any control over the power coming into your home—it's in the hands of your local utility company. But you can have a huge impact on how the electrical system inside your home performs. Whether it's knowing which breaker to turn on or off in an emergency, replacing a worn-out receptacle, installing a dimmer switch in the dining room, or wiring a new addition—all it takes is the right know-how. And that's what *Electrical Basics* is all about—know-how.

In Chapter 1, I'll explain in simple, easy-to-understand terms how your electrical system works. Then I'll take you through the components of your system and explain the purpose of each part. The second half of the chapter deals with safety. Why? Let's face it, electricity can be dangerous. Most of us have been shocked before. Whether it's accidentally sticking a finger in a light socket or carelessly unplugging a lamp, electricity can hurt you (possibly fatally). I'm not trying to scare you here. Instead, I'd like to instill in you a healthy respect for electricity. The secret to this is learning and following a set safety procedure. Once you do, you should be able to confidently tackle a wide variety of electrical projects around the home.

Chapter 2 is all about wading through the tools and often overwhelming variety of materials available for electrical work. The tools you'll need are simple—many you probably already own. All you'll need to pick up are a couple of inexpensive specialty tools, like a circuit tester and a wire stripper. Choosing materials, however, can be a numbing experience. I'll focus on the various materials you'll use most often, and what to look for when you go to the hardware store.

In Chapter 3, the fun begins: working with wire, cable, and conduit—something you'll do for every electrical job. Detailed, step-by-step instructions on everything from how to strip and join wire and cable, to running armor-clad cable and bending conduit.

Chapter 4 covers emergency repairs: dealing with blown fuses and tripped breakers and replacing faulty receptacles, cords, and plugs. Plus a section on testing switches and receptacles for proper operation.

Working with boxes, receptacles, and switches is the topic of Chapter 5. These are the meat and potatoes of electrical work. I'll walk you through selecting and installing the many types of electrical boxes, how to run cable safely in and out of the boxes, and how to replace or install receptacles and switches.

In Chapter 6 we'll look at lighting: how to replace or install ceiling fixtures, recessed lights, fluorescent lights—even how to install track lighting.

Finally, Chapter 7 delves into advanced work: adding and extending circuits such as installing a new receptacle, or running conduit along masonry walls.

Most electrical repairs and remodeling chores are not that difficult if you have the know-how. Every task in this book is broken down into discrete steps and is illustrated with detailed drawings and clear photographs. *Electrical Basics* is a step-by-step approach—a working handbook and a quick reference guide for the most common repairs you'll be faced with. I hope it helps you in your DIY endeavors.

Rick Peters
Spring 2000

Chapter 1

Home Electrical Systems

In order to work safely and confidently on your home electrical system, it's important that you have a working knowledge of the fundamentals of electricity. That's what this chapter will cover in detail. But don't worry—no formulas, calculators, or slide rules (remember those?) are required.

I'll begin by describing how a common electrical circuit works (*page 7*), using a simple analogy: the flow of water. Then on to defining the most common electrical terms you'll come across, followed by a simple description of how electricity flows into your home and through the individual circuits (*pages 8–9*).

Once you understand how electricity flows, we can look at the different parts of the electrical system in your home—from where it hooks into the local utility company's power lines at the service head, all the way to receptacles that you plug everything into (*pages 10–11*). There's also a section on how to identify the type of main service panel you have—fuse versus breaker (*pages 12–13*)—and the advantages or limitations of each type.

I've devoted the remainder of this chapter to safety, starting with the often misunderstood concepts of grounding and polarization (*pages 14–15*). Then on to what I consider one of the most important sections of the entire book: how to work safely on electrical circuits (*pages 16–19)*—everything from labeling breakers and tagging panels to working safely on energized circuits.

Finally, on *pages 20–21,* I'll walk you through the procedure on how to map the circuits in your home—that is, identifying which breakers or fuses in your main service panel control which devices (switches, receptacles, appliances, fixtures, etc.). Although this may not seem like a safety issue, it really is. In an emergency situation—an appliance shorts out and catches fire, or someone is getting shocked—knowing which breaker to turn off or which fuse to pull out can save a life.

How Circuits Work

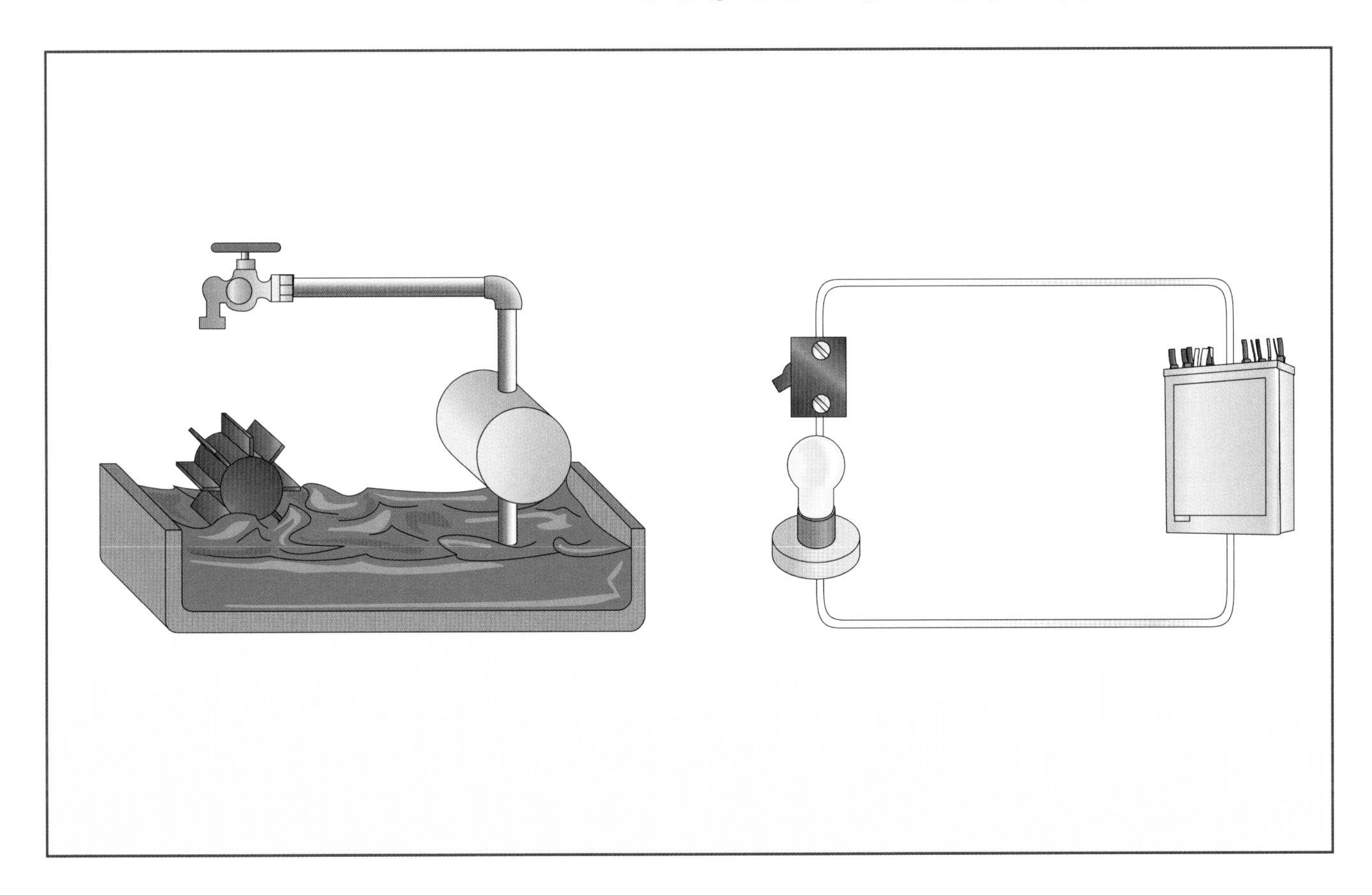

One of the simplest ways to understand how the circuits work in your home is to compare the flow of electricity to the flow of water; *see the illustration above.*

In a water-based system, a pump provides pressure to force the water to flow. Turning a faucet on or off controls the flow of the water. As the water encounters resistance in the form of a waterwheel, work is done. Then the water returns to the reservoir from which the pump draws water.

In a circuit, electrical pressure is provided by a generator (the local utility company) in the form of voltage. This voltage forces current to flow in the conductors. Current is measured in amps and describes how many electrons are flowing in a conductor at a given time. A light switch controls the current by allowing or not allowing electrons to flow. As the current encounters resistance—in the form of a lamp, for example—work is done. The current then flows back to the source (*see page 8* for the definitions of common electrical terms).

There is a well-defined relationship between the current, voltage, and resistance in any electrical circuit. This relationship is known as Ohm's law, which states that current is directly proportional to voltage and inversely proportional to resistance. In simple terms, this means when you double the voltage, the current will double. Or if you double the resistance, the current will halve. Electricians use Ohm's law all the time to calculate what type of devices and materials to use for a given job.

Common electrical terms

Alternating current (AC): the type of current found in most home electrical systems in the U.S. The current continuously varies in amplitude and periodically reverses direction or "alternates" 60 times per second (i.e., at a rate of 60 hertz).

Ampere (or amp): a unit of measure for current flow that indicates the number of electrons flowing past a point in time. 1 ampere of current is when 6.28×10^{18} electrons are flowing past a point in 1 second.

Cable: two or more wires grouped together inside a protective sheathing.

Circuit breaker: a safety device designed to protect the wiring in your home. When an overload condition exists, the breaker "trips" to safely cut off the flow of current.

Conductor: any material that allows current to flow through it; copper is an excellent conductor and is used for most wiring.

Conduit: a metal or plastic tubing that encases exposed wires to protect them.

Current: the flow of electrons past a given point in a conductor in a specified time, measured in amperes (*see Ampere, above*).

Direct current (DC): the type of current provided by a battery; unlike alternating current, direct current does not periodically change direction and is constant in amplitude.

Fuse: a safety device that monitors the flow of current; when too much current flows through a fuse, a thin metal strip melts—the fuse "blows"—to stop the flow of current.

Ground wire: a safety conductor that is part of the home wiring system: The ground wire safety shunts dangerous current to earth/ground in the event of a short circuit.

Hot wire: any conductor that brings current to a device; hot wires are covered with black or red insulation.

Insulator: a nonconductive material (usually plastic or rubber) that impedes the flow of current; conductors are encased in insulation for protection.

National Electrical Code: a body of regulations that define safe electrical procedures; local codes often add to or modify these rules.

Neutral wire: any conductor that returns current from a device to the source; neutral wires are covered with white or light gray insulation.

Overload: a demand for more current than a circuit can safely handle; an overload usually causes a circuit breaker to trip or a fuse to blow.

Power: the rate at which work is done, measured in watts; a 100-watt lightbulb consumes power at a faster rate than a 50-watt lightbulb.

Receptacle: a device designed to accept plug-in devices and provide them with power; also referred to as an outlet.

Resistance: the opposition to the flow of current, measured in ohms. A "load" (such as a lamp or radio) resists the flow of current; this resistance is what allows work to be done.

Service panel: the point where electricity from a local utility company enters the house—a metal box that distributes power to individual circuits in the home that are protected by fuses or circuit breakers.

Short circuit: a fault that occurs when two current-carrying wires make contact, or when a current-carrying conductor and a grounding conductor make contact.

Switch: a device that controls the flow of current by opening and closing the hot conductor leading to a device.

Underwriters Laboratory (UL): an organization that tests electrical devices for safety.

Voltage: the electrical "pressure" that causes current to flow in a conductor, measured in volts; 1 volt is the pressure required to move 1 amp through a conductor that has a resistance of 1 ohm.

Wattage: the amount of power a device needs or consumes; watts, the unit of measurement for electrical power, can be calculated by multiplying the voltage by the amps.

Wiring in a House

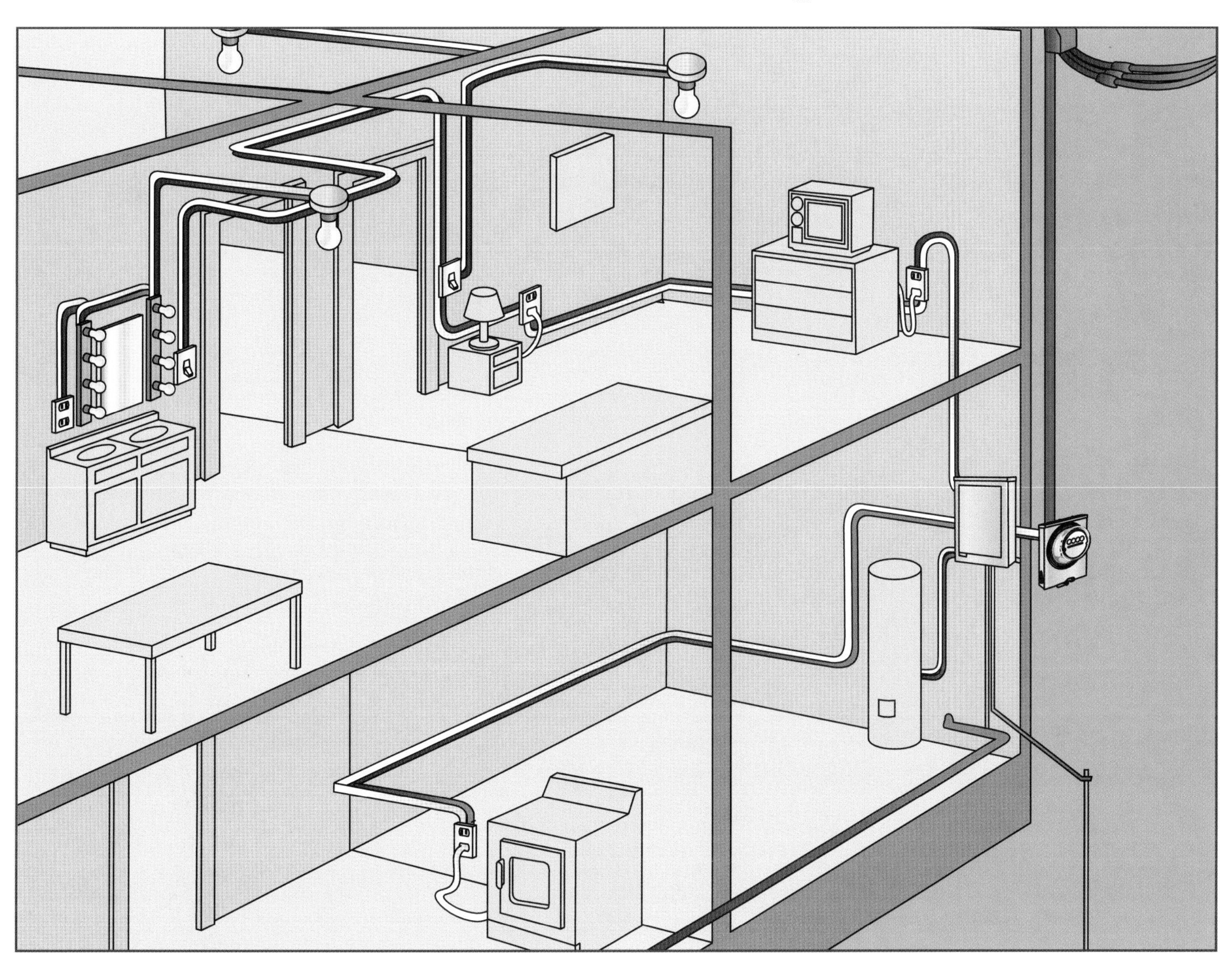

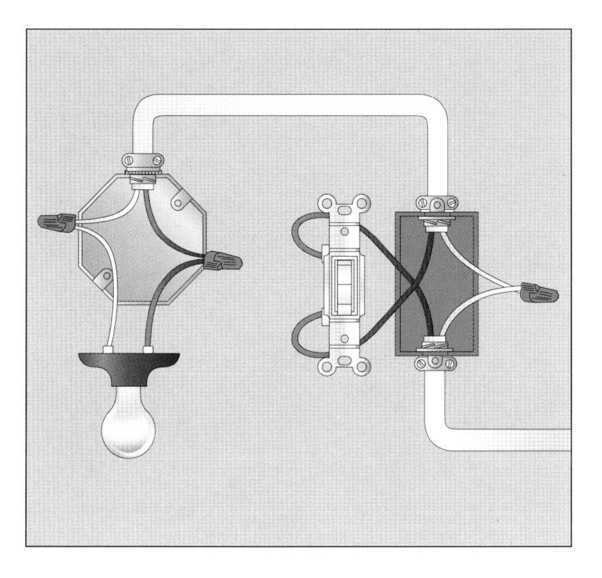

Electricity from the local utility company connects to your home through the service head. It flows through the electric company's meter and then enters the house at the service panel. Here it is distributed throughout the house via individual circuits (*see pages 10–11* for more on each part of the electrical system). Each circuit must be protected by either a fuse or a breaker. Individual circuits are connected to the service panel by way of a cable (*as illustrated at left*), or separate conductors protected by conduit. Current flows to the device (the lightbulb) through the "hot" or black wires. The "neutral" or white wires complete the path. Control devices, like switches, are always installed in the "hot" leg of the circuit.

Parts of a Residential System

Service Head Power comes into your home via a service drop from a nearby utility pole. It connects there to a service (or weather) head that anchors the wires securely to your home. The service drop, which is the responsibility of the power company, carries three wires to provide 120/240-volt operation (older homes may only have two wires, for 120-volt operation). The power company splices the service drop to your service cable—everything past the splice is your responsibility (except the meter; *see below*).

Meter The electric meter is owned by your local power company and has no user-serviceable parts inside. It's installed between the service head and the service panel to monitor the amount of power you consume. To make the meter easy to "read" by the power company, it's attached to the side of the house at a height of 2 to 6 feet, depending on the utility company. The round metal disk inside revolves as power is consumed; the faster it goes, the more power you're using.

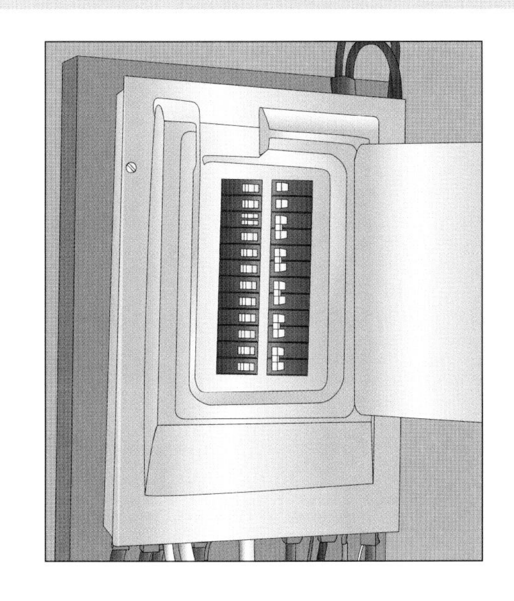

Main Service Panel Often called a breaker box or fuse box, the main service panel is usually located as close as possible to the electric meter on the opposite side, or interior wall. Its main function is to distribute power throughout the house via individual circuits. The service panel also contains devices (breakers or fuses) to protect each of the individual circuits; these devices can also be used to turn power on and off to the circuits as needed.

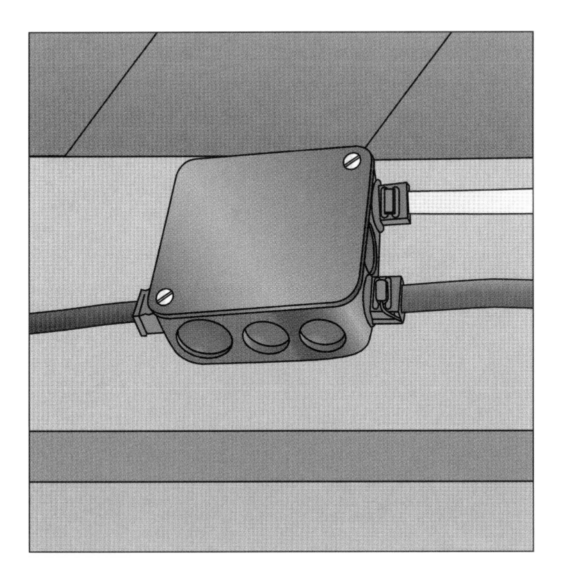

Boxes Electrical boxes contain and protect devices such as receptacles, switches, and fixtures. The National Electrical Code also requires that all splices be made in, and contained within, an approved metal or plastic box. Boxes may be inset into walls or flush-mounted to a framing member, as *shown here.* Sheathed cables (such as nonmetallic cable) can run directly into a box as long as the cable is fastened properly, uses approved wire connectors, and is not susceptible to damage.

Fixtures Most light fixtures attach directly to electrical boxes: The fixture wires are joined to the circuit and stored within the box. Other fixtures, such as recessed lights, don't require an electrical box for mounting, but they do need one nearby to encase the connections to the electrical circuit. Common light fixtures include incandescent, fluorescent, and halogen; switches control virtually all of these.

Switches and Receptacles Switches control the flow of current to devices such as light fixtures and ceiling fans. They all open the "hot" leg of the circuit and, depending of the type of switch used, can be wired to control the same device from a single location or multiple locations. Receptacles, or outlets, allow quick and safe access to the power system via any plug-in device. Special GFCI (ground-fault circuit interrupter) receptacles are now required by code to be installed in multiple locations, including kitchens and bathrooms (*see page 42*).

Service Panels

Fuse-Type

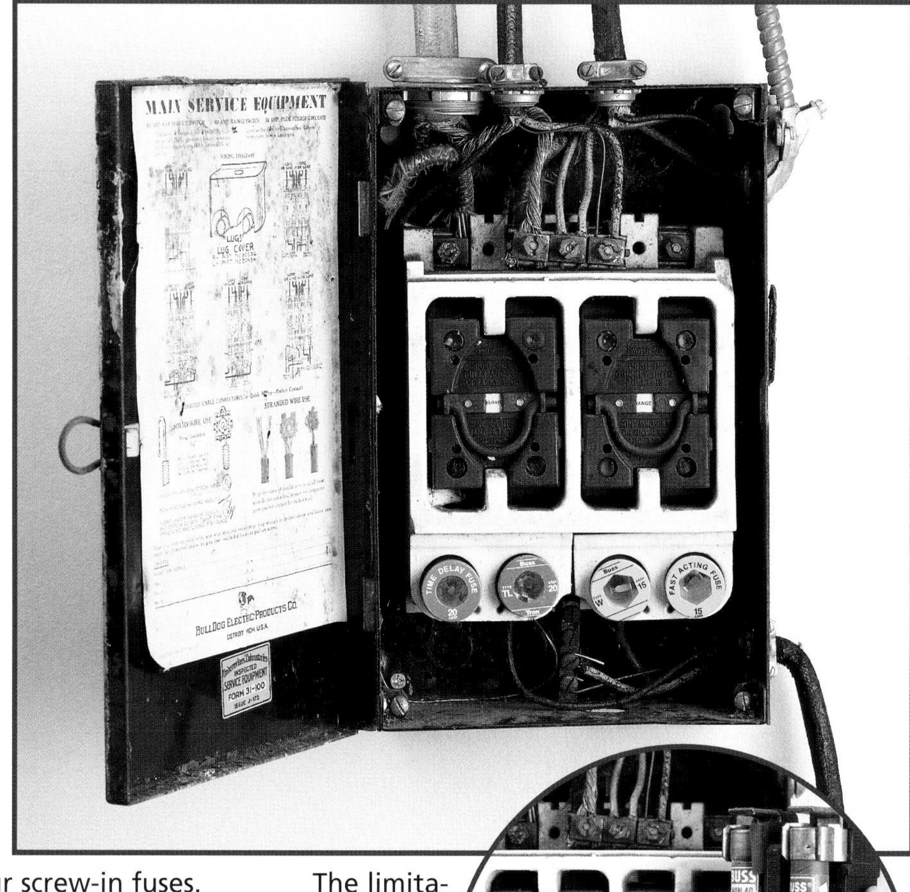

Depending on the age of your home, the main service panel will be one of two types: a fuse-based panel, or a breaker-based panel (*see page 13*). Both are designed to protect your home's wiring. Breakers and fuses are current-overload devices. Circuit current flows through them continuously; when too much current flows, the breaker "trips" or the fuse "blows" and current stops flowing. Fuses and breakers can also be used as a convenient way to turn power on and off for repairs or emergencies.

The service panel shown here is common in older homes built before 1970. It's an 60-amp service that holds two fuse blocks and four screw-in fuses. One of the fuse blocks is the main; the other is an appliance fuse block,which would typically be hooked up to an electric range. The four screw-in fuses protect individual 15-,20-, or 30-amp circuits. (Very old homes may have a 30-amp service that consists of a pair of fuses and two knife-blade–type switches.)

The screw-in fuses can be replaced by unscrewing them and installing new ones (*see page 45* for more on the different types of fuses available). Fuse blocks hold cartridge-style fuses (*see page 45*) and are removed by grasping the handle and giving it a sharp pull (*inset*). Then the cartridge fuses can be safely removed and replaced.

The limitations of an older service panel like this are readily apparent in modern homes, with our ever-increasing dependence on electricity-powered devices: TVs, VCRs, dishwashers, coffeemakers, microwaves, hair dryers, computers… the list goes on. It should come as no surprise, then, to learn that updating service panels is one of the most common (non-emergency) calls an electrician receives.

Breaker-Type

Modern service panels use circuit breakers to provide protection to a home's wiring. The big advantage of a breaker-based system is that after a current-overload condition occurs and a breaker "pops" or "trips" to remove power to the circuit, the breaker can be reset to turn power back on (*see page 69*). No more searching for fuses or frantic runs to the hardware store.

Circuit-breaker service panels started appearing in the 1960s. The most common size, 100 amps, can be found in most homes across the United States. In new construction, 200-amp services are becoming increasingly popular. Most service panels have a slot at the top for the main breaker, which shuts off power to the branch circuits, and slots for two rows of individual circuit breakers (*see page 46* for more on the different types available).

Inside the panel, there are two bus bars for connecting circuit breakers. (*inset*). The "hot" bus runs down the center of the panel, and this is where the individual breakers snap in to "tap" into the power. The neutral/grounding bus is usually located at the bottom or side of the panel. Sometimes there may be a neutral bar located on both sides of the panel. This is what the white "neutral" wires and bare copper "ground" wires attach to. The black or "hot" wires attach to screw terminals on the breakers.

Grounding and Polarization

Grounding and polarization are two of the most commonly misunderstood concepts in electrical systems. As we discussed on *page 9,* current always flows in a complete loop—it starts at the source (the service panel) and flows out through the black "hot" wire. The circuit path is completed by way of the white "neutral" wire; that is, if everything is grounded correctly and if no fault occurs.

When current flows back to the source through a path other than the neutral wire, a ground fault or "short" exists. If the short-circuited device is properly grounded, the chance of electric shock is greatly reduced. If it's not, the potential for shock is high. That's why so much attention (particularly by the code books) is concentrated on the proper grounding of a home electrical system. The whole idea behind polarization is simple. It ensures that the neutral and hot legs of a circuit are kept separate (*see page 15* for more on this).

Ground Fault on a Grounded Circuit When a ground fault or short occurs on a circuit that is grounded properly, the potentially dangerous current will flow back to the circuit panel via the equipment grounding system—the bare copper ground wire (or in the case of armor-clad cable, the outer metal housing)—and the breaker will trip, de-energizing the circuit.

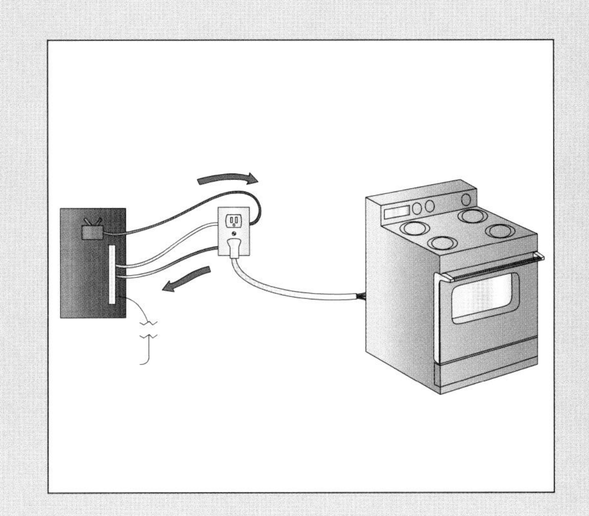

Ground Fault on an Ungrounded Circuit When a ground fault occurs on an ungrounded circuit, the current has no return path, so the breaker won't trip; no return path, that is, until someone accidentally provides one by touching the ungrounded device. Then, if the conditions permit, current will flow through the person to the ground. This is where insulated shoes and rubber mats can be lifesavers (*see page 17*).

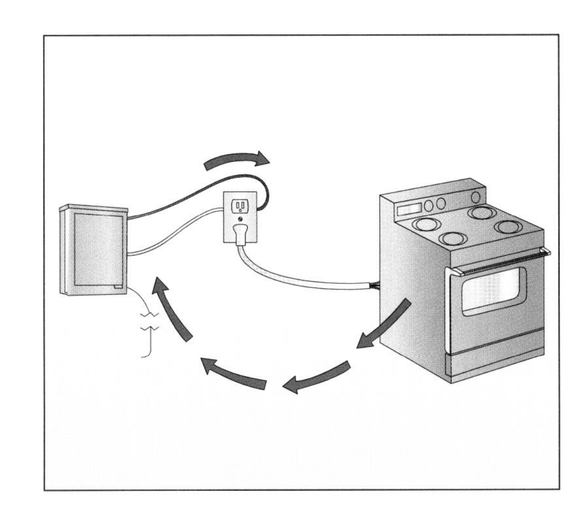

Polarized Receptacles Polarized receptacles have been around since the 1920s, when two-slot receptacles were developed. The intent of any polarized receptacle is to ensure that "hot" current flows through the black or red wires in a system and that "neutral" current flows on the white wires. The short slot is the path for "hot" current, and the long slot is the path for "neutral." Polarized receptacles that are wired correctly will ensure that polarity-sensitive devices (and there are a lot of them) will work properly.

Polarized Adapters In older homes that have receptacles without grounding holes, you can use an adapter to insert a three-prong grounded plug. These adapters are installed by first removing the receptacle cover plate screw. Then insert the adapter in the receptacle and re-install the cover plate screw. Please note that this provides ground only if the receptacle box is metal and grounded.

PREVENTING SHOCKS WITH GROUNDING

The ground portion of all the electrical circuits in your home system connect to earth/ground via a ground rod. This rod is embedded in the ground outside your home and is connected to the main service panel. How the grounding system is connected depends on the wiring in your home: Non-metallic (NM) cable uses a bare copper wire, while armor-clad cable and conduit connect via their metal housings.

Although the grounding system doesn't do anything during normal operation, it activates when a potentially dangerous situation occurs. The grounding system ensures that any metal part of a circuit you might come in contact with is connected directly to earth/ground. This prevents any harmful voltage from developing by safely shunting it to ground. When an electrical system isn't grounded properly, your body can serve as the connection between the circuit and earth/ground, resulting in a potentially fatal shock.

Working Safely with Electricity

I can't overemphasize the importance of learning to work safely with electricity. As I mentioned in the introduction to this book, what you really need to develop is a healthy respect for electricity—not a fear. Fear will only make you nervous and can lead to an accident. If you religiously follow the safety guidelines set out here, you can greatly reduce the chance of receiving a shock.

Now I'm sure some of this will seem like overkill; but trust me, it's not. One of the key elements to any safety program is developing a set routine that is always followed. Look at airline pilots—every flight begins with a walk around the plane and a visual inspection. Once inside, they go through an extensive checklist: testing controls, reading gauges, etc. Same routine every time. That's what you need to develop so that you can safely and confidently work on your home electrical system.

1 Shut off power The first thing to do whenever you need to work on a part of your electrical system is to de-energize the circuit that needs work. If you've mapped your circuits (*see pages 20–21* for instructions on how to do this), you can simply reach in and turn off the appropriate breaker or remove the desired fuse. If you haven't, you'll have to turn off each breaker, or remove each fuse in turn until the power to the desired circuit is off.

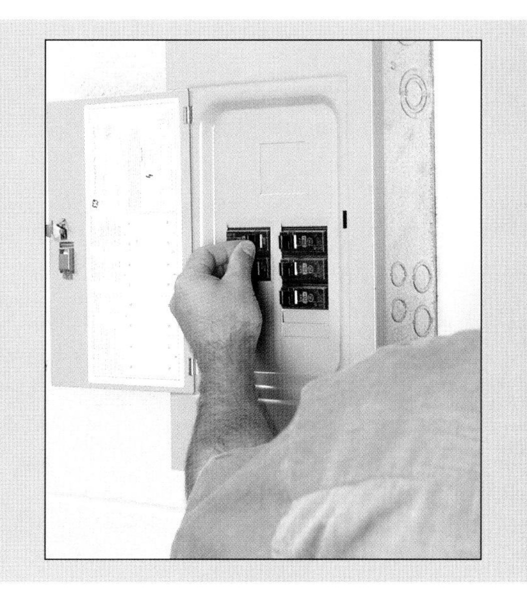

2 Tape and close panel Once you've shut off power, stick a piece of masking tape labeled *NO!* over the breaker or fuse. Then close the panel and attach a tag that warns everyone not to turn the breaker on (*inset*). These steps can prevent a potentially dangerous situation: You're working on a circuit when someone else in the house realizes they don't have power; they go to the panel and turn the breaker ON, energizing the circuit (and possibly you).

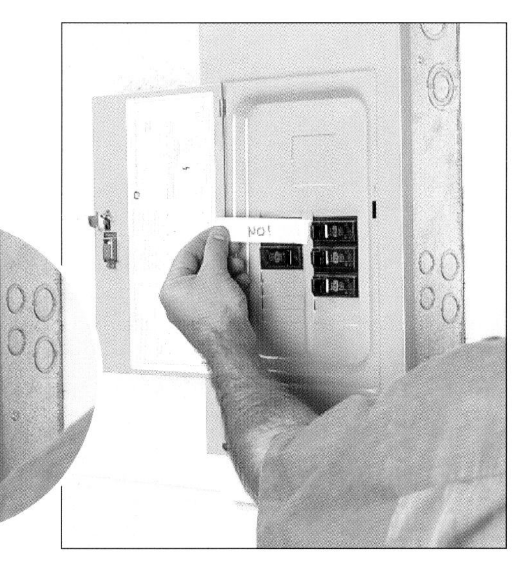

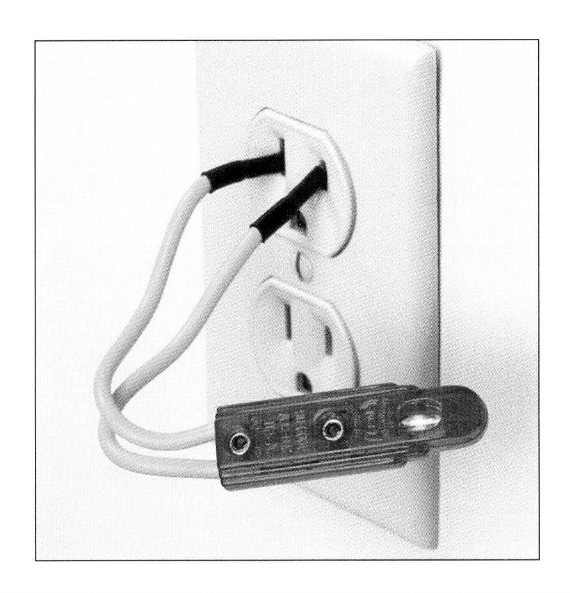

3 Check for power with a tester Even when you're positive that you've turned off the correct breaker, or removed the right fuse, you should always test the device itself for power. You can test receptacles (*as shown*) before you even remove a cover plate. On switches and light fixtures, remove the cover plate screw(s) and cover plate to gain access to the connections—check these with a circuit tester or multimeter (*see page 84*).

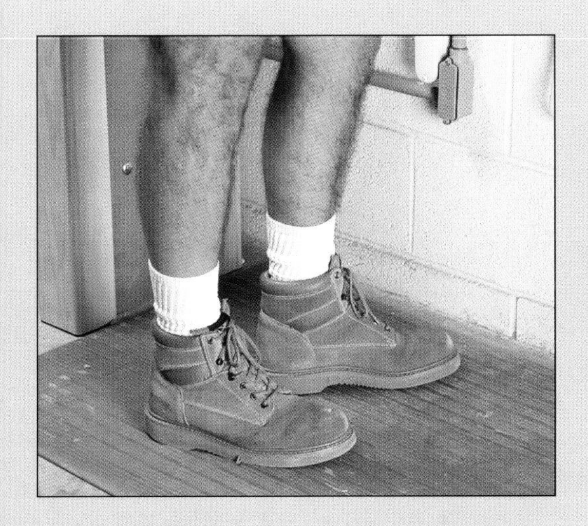

4 Wear rubber-soled shoes and/or stand on a rubber mat Whenever you need to work in or around a service panel or on an appliance or device that's encased in metal, it's a good idea to insulate yourself from ground by wearing rubber-soled shoes and/or standing on a rubber mat. This is particularly important when the floor is wet or damp. If your basement is damp like mine, I suggest you purchase a rubber floor mat and leave it on the floor directly beneath the service panel.

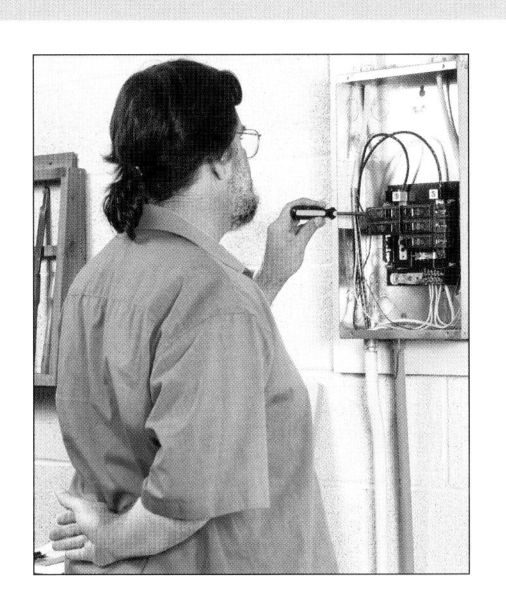

5 Keep one hand behind your back for live circuits or panels Although I advise never to work on a live circuit, an emergency may occur where you'll need to work on one (such as replacing a breaker [*see page 70*] or adding or removing wiring). Unless you've had the power company temporarily disconnect the power, there'll still be power inside the panel—even with the main breaker off. Whenever possible, don't reach inside with both hands: If you complete a circuit, the current will flow directly through your heart.

6 Don't work alone on live circuits If you do have to work on a live circuit, don't work alone. If you accidentally bridge the circuit and get shocked, you most likely won't be able to disconnect yourself. Unlike direct current (DC), which tends to repel a person when they're shocked, alternating current (AC) has a tendency to "grab" a person. A helper armed with a rope or broom handle can pull or knock you safely away from the circuit (if they simply grab you, they'll get shocked, too).

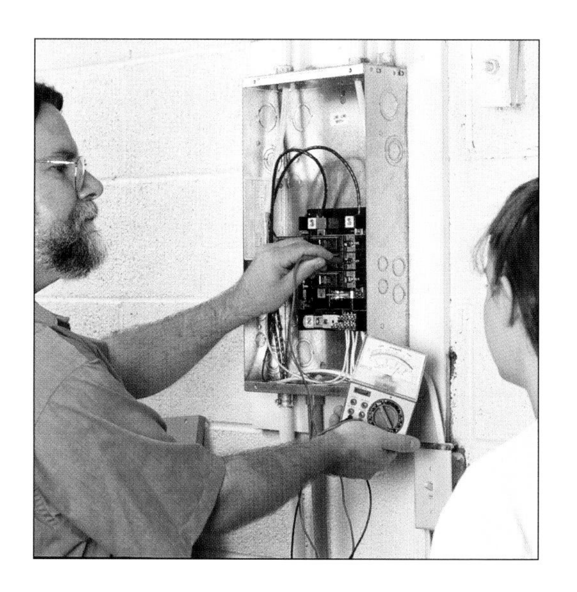

7 Don't touch metal This rule is an extension of the "work with one hand behind the back" rule. Whenever possible, don't touch metal. There may be times when you're working on a main service panel or metal-encased appliance and you slip or lose your balance. If this happens, make a conscious effort not to touch any metal: Reach for siding, masonry, plastic—anything but metal—to prevent the possibility of bridging a circuit and receiving a nasty shock.

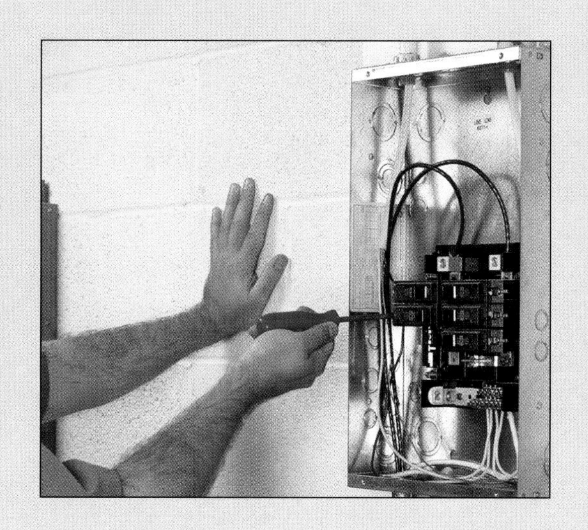

8 Never alter safety devices This is a photo of a switch box for a water heater that a neighbor asked me to look at. When the fuses blew on the previous owners, they were too lazy to replace them. So instead, they bypassed the fuses by joining the incoming and outgoing lines. Don't ever alter a safety device! My neighbor was lucky that the water heater never malfunctioned. If it had, there was no circuit protection. The wires would have overheated and could have easily started a fire.

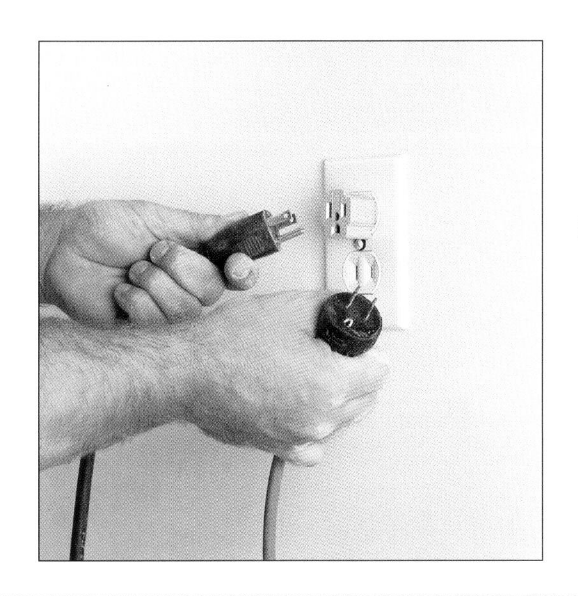

9 Don't alter plugs Altering a plug to fit an outlet is like playing Russian roulette: Eventually someone is going to get hurt. Don't be tempted to break off a grounding prong of a grounded plug so that it'll fit into a two-slot receptacle. That prong is there to protect you. An alternative to this is to use an adapter that's held in place with a cover plate screw. Note: An adapter like this will afford ground protection only if the receptacle box is metal and is grounded properly.

10 Use UL-approved parts Another way that you can protect yourself and your loved ones is to use only UL (Underwriters Laboratory) listed parts for your electrical projects. (The Underwriters Laboratory is an organization that tests electrical devices for safety.) Check each part and look for the UL label before you buy it.

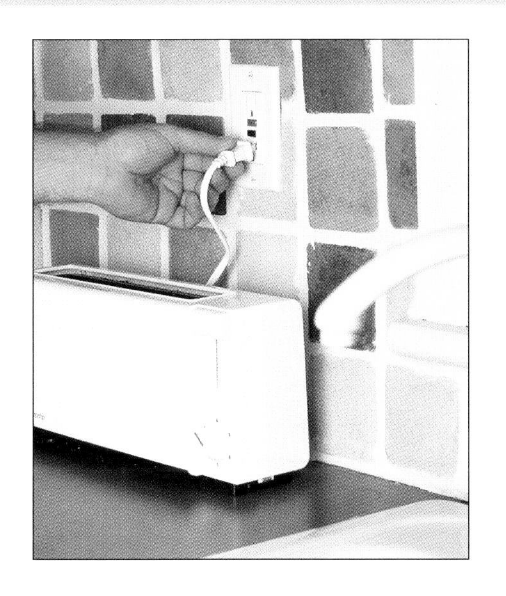

11 Use GFCI outlets You can also protect your family from potentially dangerous shock hazards by installing GFCI (ground-fault circuit interrupter) receptacles wherever they're called for by your local code—typically in bathrooms, kitchens, pool areas, garages, and outdoors. These devices monitor current flow and cut off power when it begins to flow where it's not supposed to (*see page 42* for more on these lifesaving devices).

Mapping Your Circuits

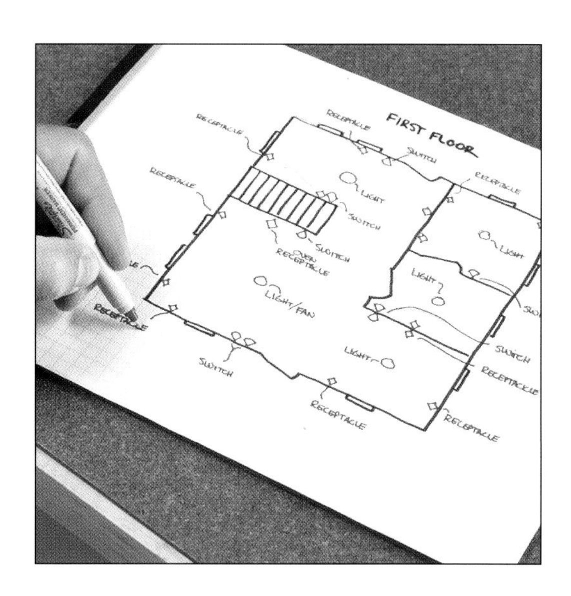

1 Sketch the rooms, label the fixtures Mapping your circuits can save you a lot of time and frustration fumbling around in the main service panel, turning off breakers or pulling fuses. Knowing which fuse or breaker controls the different devices in your home can even save a life in an emergency. Start by drawing a rough sketch of your floor plan, one for each floor. Note where each switch, receptacle, fixture, and appliance is on the drawing. Include windows, doors, and stairs for reference.

2 Label breakers or fuses Now go to the main service panel and place a numbered label on each breaker or fuse. Once you've determined which breaker or fuse controls which devices, you'll write this number on the drawing you made in Step 1 (*see Step 5*).

3 Turn on one at a time With the breakers labeled, you can start testing the individual circuits. To do this, begin by turning off all the breakers, or unscrewing all the fuses. Leave the main breaker on, or the main fuse in. Now flip on one circuit breaker at a time, or screw in one fuse at a time.

4 Test each outlet/fixture A helper is useful for this job, but you can do it yourself. Start by turning on switches and lights, and check appliances throughout the house: Systematically insert a circuit tester (or lamp or radio in the ON position) in each outlet. Identify any device that has power. Temporarily stick a label or piece of masking tape on each powered device, and note the breaker or fuse number and the circuit's current rating (as indicated on the fuse or breaker).

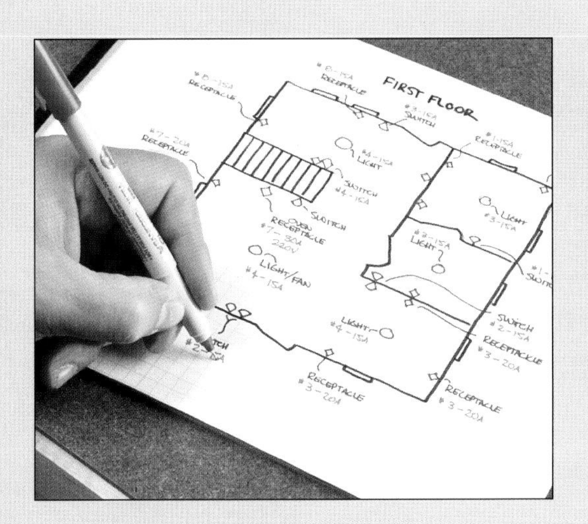

5 Label the drawing from Step 1 After you've turned on each breaker or fuse and have checked the entire house for power, make sure every device has a label. If you find one without a label, have your helper turn each circuit off in turn until you identify its breaker or fuse. Then walk around the house and transfer the information from the labels you put on the devices in Step 4 onto the floor plan drawing you made in Step 1.

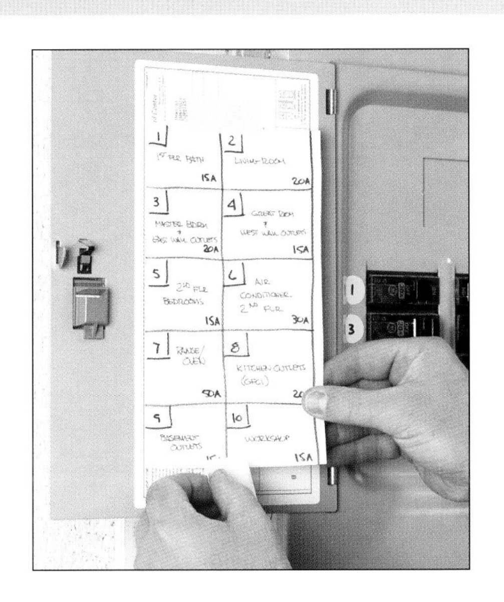

6 Apply index to panel and keep drawings nearby Draw an index of the breakers or fuses in your service panel and label each with a brief description of what the breaker or fuse controls—hot water heater, receptacles in living room, etc.—and tape it inside the service panel door. Then affix the drawing from Step 5 to the service panel or to a nearby wall. Now the next time a repair needs to be made, or an emergency pops up, you'll know exactly which fuse or breaker to reach for.

Chapter 2

Electrical Tools and Materials

Two of the most important steps in successfully completing an electrical project safely are using the right tools and selecting the proper materials. The tool part is pretty easy—there are only a few specialty tools you'll need, and many of the other tools you probably already have in your toolbox. The tougher task is selecting materials.

There are a couple reasons for this. First, if you've ever wandered up and down the electrical aisle at a warehouse-type home center, you know that there is a dizzying array of materials to choose from—and many of them appear identical (switches and receptacles, just to name a few). Second, many of the materials you can buy may not be code-approved for installation in your home.

Why do some stores sell non-code-approved parts, you ask? Because many national chains often stock the same materials from store to store—they can't afford to customize their aisle offerings to meet the local electrical code (which varies from town to town and from state to state). This is an excellent reason to purchase your materials from the local electrical supply house: They usually don't sell stuff that's not approved for local installation, because they cater to the area electricians; if they did, they'd never get any return business. You may have to pay a bit more at an electrical supply house, but you'll get quality parts (often heavy-duty, commercial-grade) that will meet code.

In this chapter, I'll start by describing the tools you'll need to tackle most electrical jobs around the house: general-purpose tools (such as gripping tools, power tools, etc.) on *pages 23–24,* and specialty tools (like circuit testers, wire strippers, and conduit benders) on *page 25.*

Then on to the myriad materials available. What to look for when you go to buy: wire and cable (*pages 26–29*); conduit and connectors (*pages 30–33*); electrical boxes (*pages 34–35*); switches (*pages 36–39*); receptacles and cover plates (*pages 40–43*); connectors and restraints (*page 44*); and fuses, circuit breakers, and service panels (*pages 45, 46, and 47, respectively*).

General-Purpose Tools

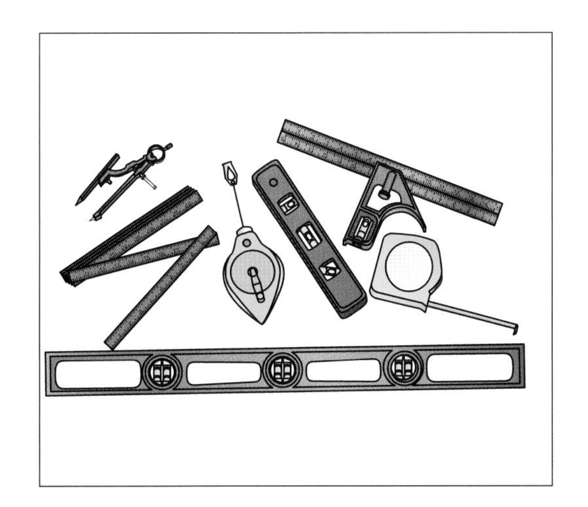

Measuring/Layout One of the most critical steps in any electrical remodeling job is measuring and laying out the placement of fixtures. The tools *shown* should be in every homeowner's toolbox (*clockwise from center*): a 3' level and a torpedo level; a 25' tape measure; a combination square to check for right angles and for general layout; a chalk line for striking long layout lines on walls and floors; a compass to draw circles; and a folding rule for short accurate measurements.

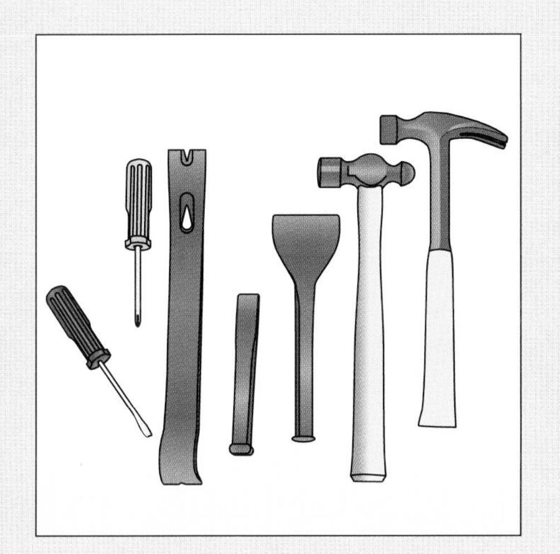

Demolition Many of the electrical jobs you'll tackle will require some demolition work; that is, tearing out an old section of wall, flooring, or cabinet. You'll find the following tools useful for this type of work (*from left to right*): screwdrivers for general dismantling; a pry bar for pulling out stubborn boards and fixtures; a cold chisel or set of inexpensive chisels for chopping out holes in walls or flooring; and a ball peen hammer or claw hammer.

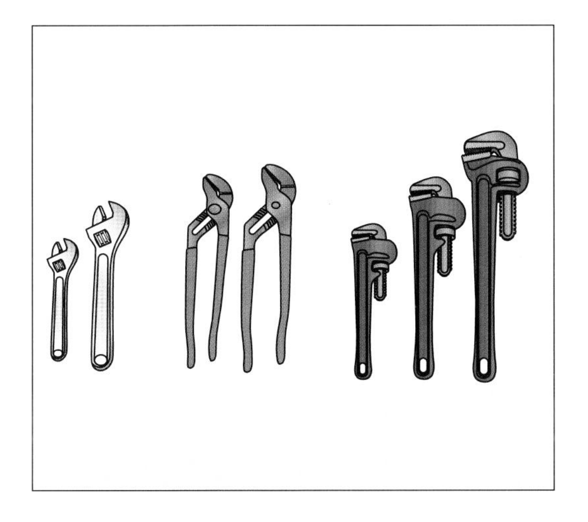

Gripping Electrical work requires a wide variety of gripping tools for loosening and tightening nuts, caps, fittings, etc. Although it's helpful to have different sizes, you should have at least one of each of the following tools in your toolbox (*from left to right*): an adjustable wrench for smaller nuts and fittings; a set of channel-type pliers for larger nuts and fittings; and a pipe wrench for working with threaded pipe.

Cutting Tools Many electrical tasks will require you to cut a wide variety of materials. The cutting tools to have handy are (*counterclockwise from top left*): a drywall saw for "cutting in" electrical boxes; a compass saw for cutting notches in framing members, flooring, cabinets, etc.; a pocketknife or utility knife to trim away insulation and cut sheathing; and needle-nose pliers and diagonal cutters for trimming wires and cutting cable.

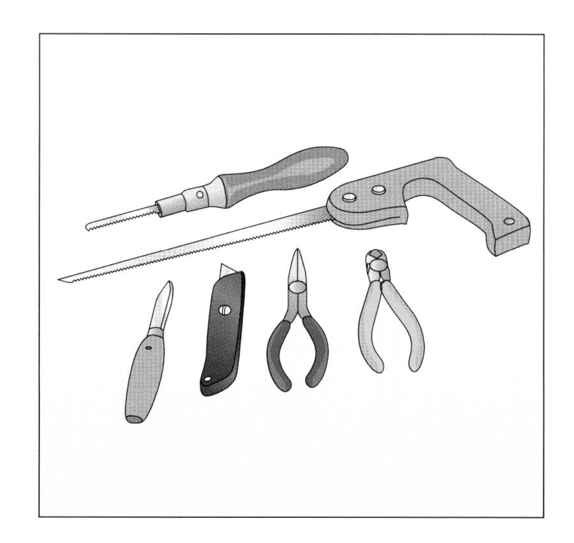

Power Tools Used primarily for electrical remodeling, power tools can make quick work of many tedious jobs. *Clockwise from top left:* a cordless trim saw for straight-square cuts; a saber saw for cutting access holes; a cordless drill with a 3/8" chuck for smaller-diameter holes; a right-angle drill for tight spots; an electric drill with a 1/2" chuck for larger-diameter holes; and a reciprocating saw for cutting through walls and cabinets.

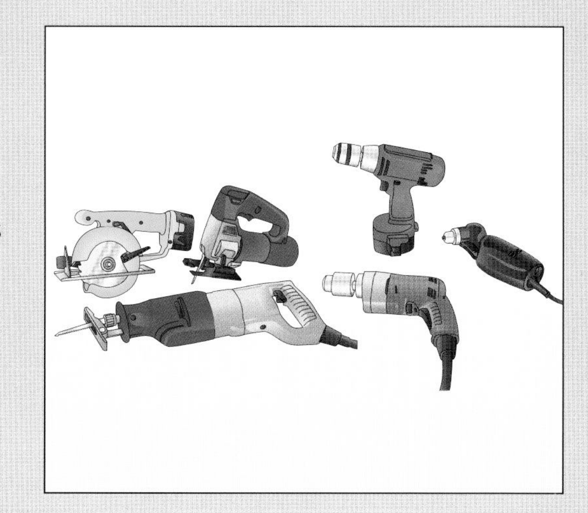

Protective Gear As with any home improvement job, it's important to protect yourself by wearing appropriate protective gear. Keep the following on hand (*clockwise from bottom left*): leather gloves to protect your hands; knee pads not only to cushion your knees but also to protect them; ear muffs or plugs when working with power tools; rubber gloves for added protection against shocks; a fuse puller to safely remove cartridge fuses; and safety goggles to protect your eyes.

Specialty Tools

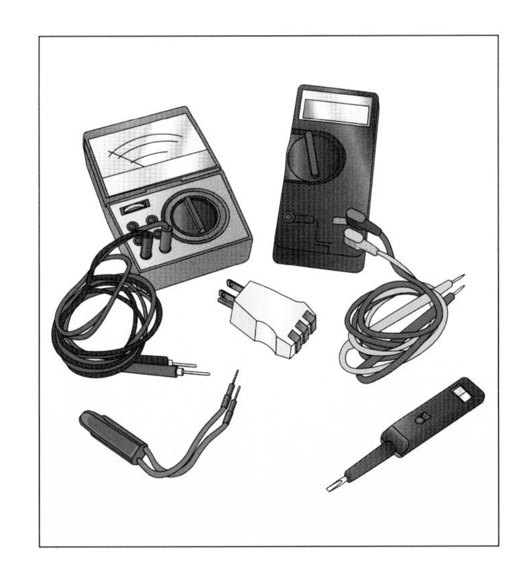

Electrical Testers Critical to all electrical work is the ability to test for power. *Shown* are (*clockwise from top left*): an analog multimeter (capable of measuring voltage, amperage, and resistance); a digital auto-ranging multimeter (it automatically selects the most appropriate range); a plug-in circuit analyzer *(center)* that reveals the condition of a receptacle's wiring; and a probe and a neon circuit tester—both show the absence or presence of power.

Wire-Working Tools At the heart of every electrician's toolbox are the tools for working wire and cable. *Shown* are (*counterclockwise from top left*): a cable ripper for removing the sheathing from non-metallic cable; a combination wire stripper/crimping tool for stripping wire and securing crimp-type connectors to wire; linesman pliers—the most reached-for tool in a tool belt, since it can both cut and grip; a pair of wire strippers with notches sized for various-gauge wire; and a pair of nippers for cutting wire flush with a surface.

Conduit Tools If you're planning on working with conduit, you'll need some specialty tools. *Clockwise from top left:* a conduit bender for working conduit around corners; wire-pulling lubricant to help wire slide smoothly through a long run; a hacksaw for cutting conduit; a fish tape to pull wires through conduit (and walls); electrician's tape for splicing wires to the fish tape; and a standard tubing cutter to cut conduit clean and square.

Types of Wire

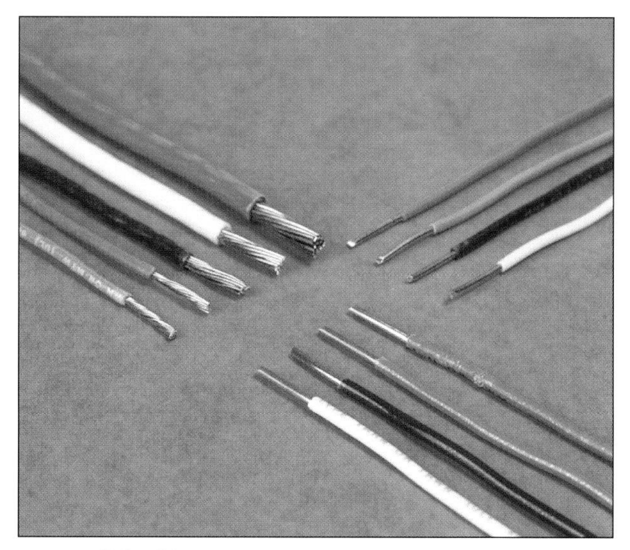

There are two main types of copper wire that you'll use for electrical work: solid and stranded. (Occasionally you may come across aluminum wire; *see the sidebar on the opposite page.*) Electrical wire is sized according to the American Wire Gauge (AWG) system; *see the chart on the opposite page.*

Basically, the smaller the number, the larger the wire. Working with the larger sizes (smaller gauges) of copper wire can be difficult. That's where the more flexible stranded wire comes into play. It's a lot easier to handle, but not as simple to terminate because of the multiple strands.

Pictured in the photo above right are common wires used for electrical projects. All of those shown are type THHN/THWN: They're covered with a thermoplastic insulating jacket and have excellent resistance to heat and moisture. *Clockwise from top left:* stranded wire (*top to bottom*)—6-gauge (green), 8-gauge (white), 10-gauge (black), 12-gauge (red), 14-gauge (green); in solid wire: 14-gauge in red, green, black, and white, and 12-gauge in the same colors.

The two sizes that you'll be using the most in your home are 12- and 14-gauge wire. They're designed to handle 20 and 15 amps, respectively. The different wire colors indicate the intended use of the conductor; *see the chart below.* For your safety and the safety of anyone working on your electrical system, it's imperative that you follow this color standard.

	White	**Black**	**Red**	**White with Black Markings**	**Green**	**Bare Copper**
FUNCTION	A neutral wire or "return" that transports current to the source at 0 voltage.	A "hot" wire that transports current to a device at full voltage.	A "hot" wire often referred to as a "control" for switching that transports current at full voltage.	A "hot" wire often referred to as a "control" for switching that transports current at full voltage.	An insulated conductor that serves as a path for grounding.	A bare conductor that serves as a path for grounding.

Wire gauge capacity and usage

	Wire Guage						
	6	8	10	12	14	16	18
Capacity	60 amps 240 volts	40 amps 240 volts	30 amps 240 volts	20 amps 120 volts	15 amps 120 volts		
Usage	Electric furnace, central air condition-ing	Central air condi-tioning, electric range	Clothes dryer, window air conditioner	Recep-tacles, switches, light fixtures	Recep-tacles, switches, light fixtures	Extension cords	Low-voltage lighting, "zip" cords for lamps

ALUMINUM WIRE

Prior to the 1970s, aluminum wire was used in some homes in place of copper (copper prices were high at the time). In most cases, its silver color will identify it immediately as aluminum wire. If not, check the sheathing or insulation for an AL or aluminum manufacturer's stamp. In the early 1970s, it was determined that aluminum wiring presented a safety hazard in the home: The wire expands and contracts differently than copper and could work loose from screw terminals over time.

If your home is wired with aluminum, it's a good idea to call in an electrical inspector to make sure that it's safe. As long as it's installed properly, aluminum wire is safe. In fact, it's still widely used today to bring power into the home via the service mast.

Working with aluminum wire requires special handling. First, smaller-gauge wire is brittle and breaks easily. Second, once the insulation is removed, the conductors must be coated with an anti-oxidation compound to prevent oxidation, which can lead to problems. Since this type of wire has special installation and code requirements, I'd advise calling in an electrician if you need to work with it.

Types of Cable

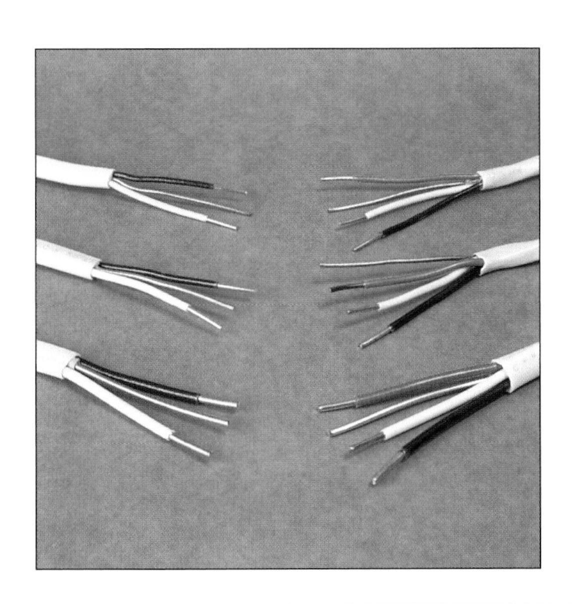

NM (Non-Metallic) Non-metallic cable, often referred to as Romex (a registered trademark of the General Cable Company), consists of two or more insulated conductors in a single non-metallic jacket or sheathing. The conductors are wrapped with a paper insulation before they're surrounded with the sheathing. The NM cables *shown here* are the types you'll use most often (*clockwise from bottom left*): 10/2, 12/2, 14/2, 14/3, 12/3, and 10/3.

Old NM/Braided Although many folks believe that NM cable is a modern invention, it's been around for decades. Early NM cable (the type used in the 1950s) consisted of a pair of conductors surrounded by a braided jacket, usually brown or silver-colored. The braided insulation often degrades over time (especially if it gets hot); and as you can see, it's not the tidiest stuff to work with.

READING NM CABLE

Non-metallic cable is usually labeled by a manufacturer to identify the number of conductors and their wire gauge. (It's important to note that the bare copper grounding wire is not counted.) The first number (reading from left to right) indicates the gauge of the conductors, and the second number (following a slash) describes the number of conductors. The top cable *shown*, 14/3 with ground, uses 14-gauge wire with black, white, and red conductors—along with a bare wire for grounding. The bottom cable, 12/2 with ground, is 12-gauge wire with two conductors.

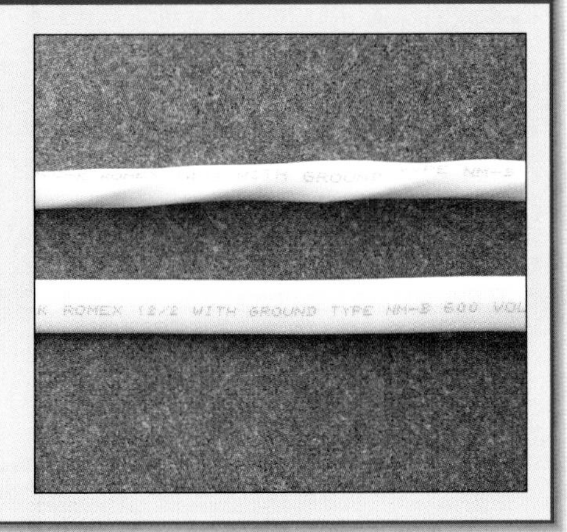

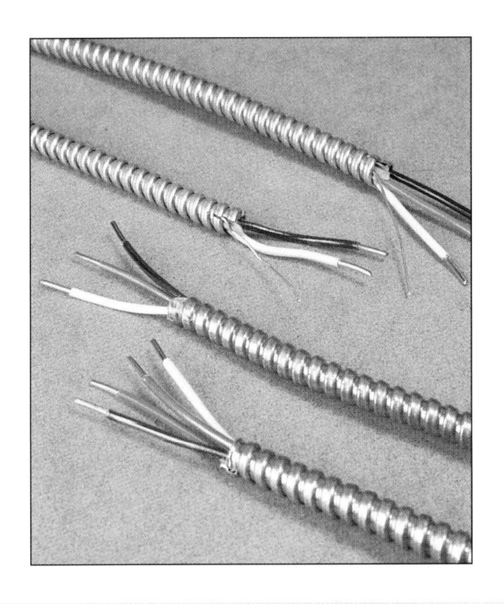

Armor-Clad Cable Armor-clad or BX cable (a registered trademark of General Electric), holds two to four insulated conductors. BX also has an aluminum bonding wire that makes contact with the armor along its length and is used as part of the grounding system. It is not, however, intended for use as a grounding wire; the armor itself serves as the ground. MX is armor-clad cable without a bonding wire. *Shown, from top to bottom:* 10/3 BX, 10/2 BX, 10/2 MX, and 10/3 MX.

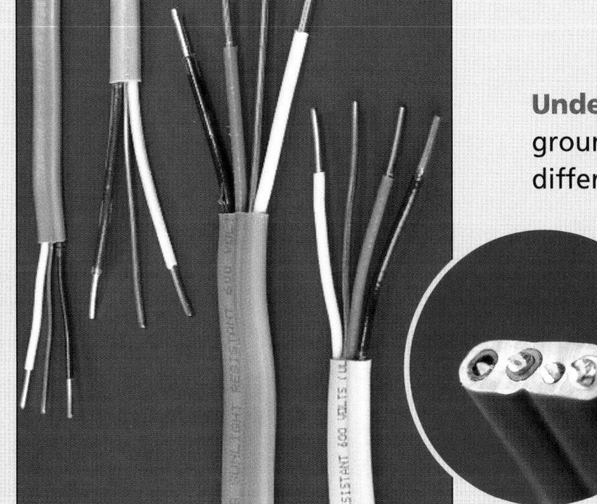

Underground Cable Underground cable, or UF (underground feeder), is similar to ordinary NM cable. The big difference is that the conductors in UF cable are encased in thermoplastic (*inset*). Because of this, underground cable can be buried directly in the ground (check you local code for trench size, location, depth, etc.). Underground feeder cable is used when moisture is a concern, such as running an underground circuit to an outdoor light.

SELECTING THE RIGHT CABLE

Which type of cable is best for the job at hand? It all depends on your local code and on where the cable is being run. Non-metallic (NM) cable is approved in most localities for use inside floors, walls, and other areas where it can't get damaged or wet. It's your best bet in most cases. If you need to run a short length of cable in an exposed or damp area, many codes will allow you to use armor-clad cable. If not, you'll most likely need to run conduit. If you want to run cable underground without installing conduit (check your local code to see whether this is approved), use UF cable. Keep in mind, however, that underground cable must never be spliced—it must be a continuous run from the power source to the fixture.

Conduit and Connectors

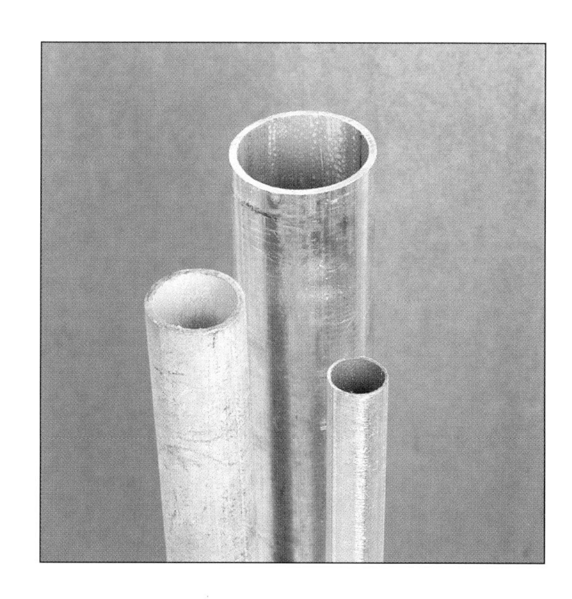

Conduit Exposed wiring in some locations must be protected with conduit (check your local code). There are three types (*clockwise from bottom right*): EMT, IMC, and rigid. EMT is the most common; it's lightweight and easy to use. Thicker-walled IMC offers better protection. But for the ultimate, use rigid conduit (note that it's expensive and requires threaded fittings). All conduit is rated to carry a certain number of conductors; *see the chart below.*

Couplings and Elbows and Offsets To make conduit easy to work with, a variety of shaped fittings are available. Elbow fittings (the four fittings on the *far right of the photo*) let you go around corners without bending pipe. Box connectors (top two *on left*) join the conduit to a box. Couplings (*middle left*) join sections of conduit together using either setscrews or compression fittings. And an offset (*bottom left*) makes a smooth transition from the electrical box to the wall.

Allowable conductors in EMT conduit

	no. of 14-gauge conductors	no. of 12-gauge conductors	no. of 10-gauge conductors	no. of 8-gauge conductors	no. of 6-gauge conductors	no. of 4-gauge conductors
½" EMT conduit	12	9	5	3	2	1
¾" EMT conduit	22	16	10	6	4	2
1" EMT conduit	35	26	16	9	7	4
1½" EMT conduit	84	61	38	22	16	10

Cable Connectors In addition to connectors that join conduit to metal boxes, there are other connectors available designed to connect virtually any kind of cable to an electrical box. *Clockwise from top left:* a 90-degree armor-clad connector that clamps the cable in place with a pair of setscrews; a straight, screw-in connector for armor-clad cable; and a "universal" connector that can accept a variety of cables, including non-metallic cable.

Single/Double Boxes and Ganged Boxes The box you choose for a project will depend primarily on the number of conductors it needs to hold (*see the chart below*), along with what will be installed in the box (switch, receptacle, etc.). Some common boxes you'll use are (*counterclockwise from top left*): a 4"-square, 1½"-deep box; a 3"×2" rectangular box, 2½" deep; and a "ganged" box, made by taking two single boxes of the same depth, removing a "wall" on each, and screwing them together.

Allowable conductors in an electrical box

	no. of 18-gauge conductors	no. of 16-gauge conductors	no. of 14-gauge conductors	no. of 12-gauge conductors	no. of 10-gauge conductors	no. of 8-gauge conductors
4" square, 1½" deep	14	12	10	9	8	7
4" square, 2⅛" deep	20	17	15	13	12	10
4" octagonal, 1½" deep	10	8	7	6	6	5
4" octagonal, 2⅛" deep	14	12	10	9	8	7
3"×2" rectangular, 1½" deep	5	4	3	3	3	2
3"×2" rectangular, 2¼" deep	7	6	5	4	4	3
3"×2" rectangular, 3½" deep	12	10	9	8	7	6

Utility Boxes Although utility boxes can be recessed in walls and ceilings just like standard boxes, they differ from standard boxes in that their corners are rounded. This makes them a better choice when the boxes will be exposed, such as in a garage or basement with masonry walls. Removing the sharp corners both improves appearance and reduces the risk of injury. *Clockwise from top:* a single box, an octagon, and a double box.

Ceiling Boxes The type of ceiling box you choose will depend on what it has to support. Plastic boxes should be used only for lightweight fixtures. Use metal boxes and brackets for heavy fixtures—ceiling fans require a special box rated to support their weight. *Clockwise from top left:* a round plastic nail-on box; a round plastic box with bracket; a metal octagonal box with bracket; a metal octagonal box for cut-ins; and, *center,* a round plastic box with sliding bracket.

Thin-Wall Boxes Thin-wall boxes are a special type of box designed for installations where wall thickness is a problem, such as paneled walls in a basement. They come in different depths and in both metal and plastic, and they can be much thinner than their standard cousins (*shown here behind the thin-wall boxes*). Since the reduced depth decreases the number of conductors and devices each box can hold, check your local code to see whether these boxes are approved for use.

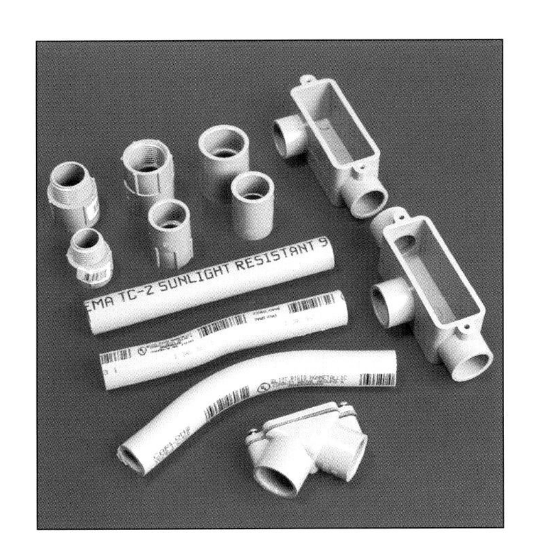

Plastic Conduit and Boxes

Conduit Plastic conduit is gaining in popularity for good reason: It's inexpensive and nonconductive. (*See the chart below* for allowable conductors.) *Clockwise from top right:* a 90-degree condulet, a T-condulet, a pulling elbow, a 45-degree bend, an offset, ½" conduit; and *on top from left to right* in ½" and ¾": male threaded connectors to join conduit to a box, female threaded connectors, and couplings for joining together sections of conduit.

Boxes Regardless of what size or shape PVC box that you choose, the box will have either threaded openings (*right*) or smooth openings (*left*) for cementing conduit in place. Plastic boxes are not without some limitations. First, they're not as versatile as metal boxes since they can't be "ganged" together (*see page 31*). Second, they're not as strong or sturdy as metal boxes. And finally, they can't be used if you're working with metal conduit or armor-clad cable.

Allowable conductors in PVC conduit

	no. of 14-gauge conductors	no. of 12-gauge conductors	no. of 10-gauge conductors	no. of 8-gauge conductors	no. of 6-gauge conductors	no. of 4-gauge conductors
½" PVC conduit	**10**	**7**	**4**	**2**	**1**	**1**
¾" PVC conduit	**18**	**13**	**8**	**5**	**3**	**1**
1" PVC conduit	**32**	**23**	**15**	**8**	**6**	**4**
1½" PVC conduit	**80**	**58**	**36**	**21**	**14**	**9**

Boxes – New Construction

Electrical boxes for new construction (or for spaces where the framing is exposed, like an unfinished basement or garage) are designed for ease and speed of installation, since time is money on the construction site.

First, virtually every style of new-construction box has a built-in gauge to make it easy to align the box and install it so the outer lip of the box will end up flush with the finished wall surface. Second, many of the new construction–box manufacturers have provided built-in brackets or fastening systems (nails, tabs, etc.) that require no additional fasteners: Just grab a hammer and go.

Metal or Plastic New-construction boxes are either metal or plastic. *Clockwise from top left:* a 2"×4" plastic nail-on box, a 2"×4" plastic box with a front nailing bracket, a 4"-square metal box with clip bracket, a 2"×4" metal box with side nailing bracket, and a 2"×4" plastic box with a side nailing bracket. Note: Most metal brackets have tabs that you can drive into the framing to hold the box in place until you can secure it with nails or screws.

Adjustable A relative newcomer to the trade is the plastic adjustable box. This type of box looks just like a standard box with a side bracket. But on closer examination, you'll see a depth-adjustment screw that runs alongside the bracket. Turning the screw moves the entire box in and out with respect to the bracket so that you can adjust it flush with any type of wall surface. These are particularly useful when working with tiled or paneled walls.

Boxes – Existing Construction

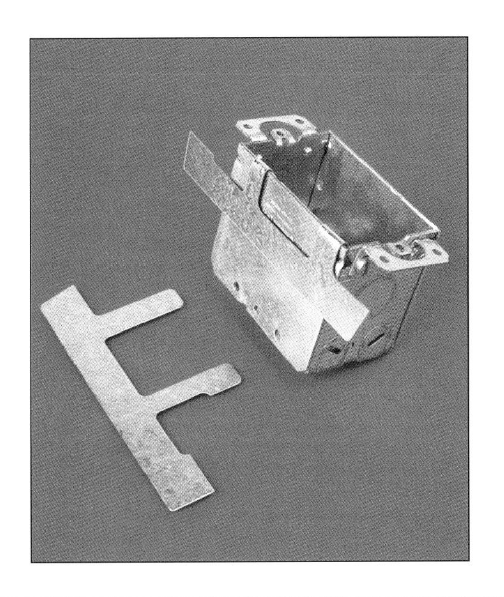

Metal Straps Any box that has drywall ears (the small L-shaped brackets on its top and bottom) can be installed in an existing wall by using a pair of metal straps. Once a hole is cut in the wall, insert the box. Then slip the thin metal straps (one on each side of the box) between the box and the opening. While holding the box in place, pull the strap forward as far as possible and fold in the metal "arms" or "tabs" (*as shown*) to support the box.

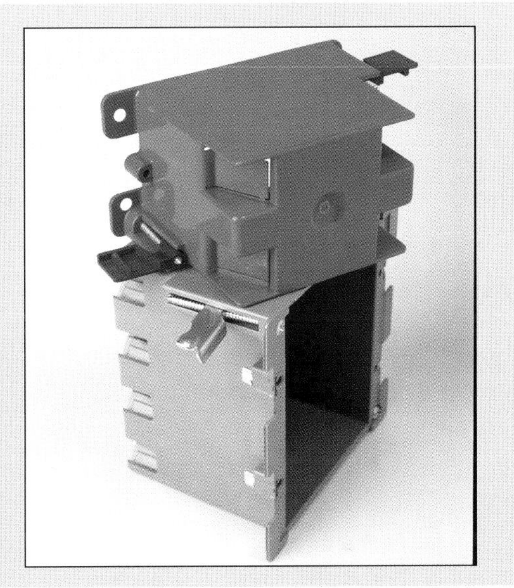

Plastic or Metal "Ears" Another box that can be "cut in" to an existing wall uses plastic "ears" to clamp the box to the wall. After a hole is cut in the wall (most manufacturers of this style box provide a cutout template), press the box into place. The ears will flip into position as soon as you begin tightening the screws. As you continue to tighten each screw, the ears will pull the box tight up against the wall.

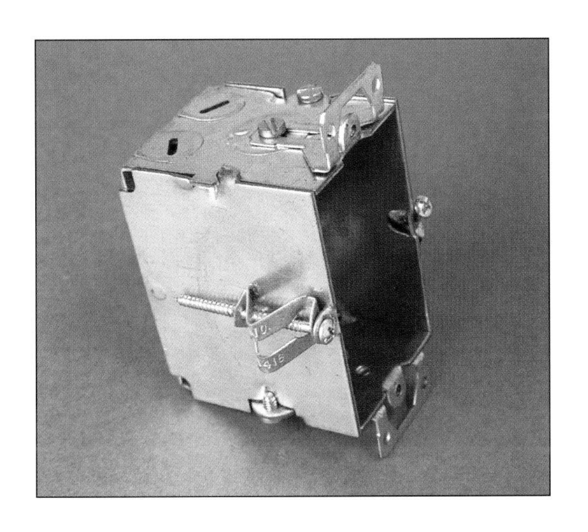

Clamp Type One of the sturdiest "cut in" boxes you can use is a metal clamp type. This style of box incorporates a pair of "toggle bolt"–style clamps that expand *as shown* to secure the box to a wall. The hole for this style of box needs two additional notches in the sides to provide clearance for the clamps. If the manufacturer doesn't provide a template, make your own by placing the box face-down on a sheet of paper and tracing around it.

Switches

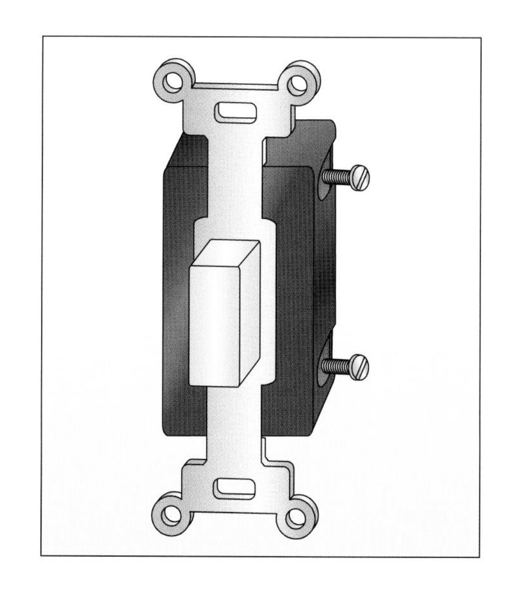

Single-Pole A single-pole switch is the most common type of switch. It controls a light fixture or receptacle from a single location. It is easily identified by its two brass-colored screw terminals on one side (some switches also have a green screw terminal for ground). Most single-pole switches are clearly labeled ON and OFF on the switch lever; the push-button-style switch shown is an exception. (*See page 98* for how these switches are wired.)

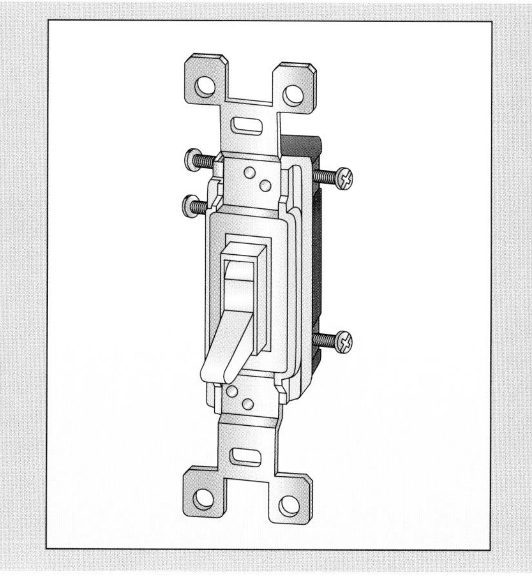

Three-Way Three-way switches control a light fixture or receptacle from two locations. They are distinguishable from a single-pole switch in two ways. First, you'll find three screw terminals instead of two—the two like-colored screws are "travelers," and the darker (or black) terminal is the common. Second, since these switches can be up or down for ON or OFF, the lever has no markings. (*See page 99* for wiring instructions.)

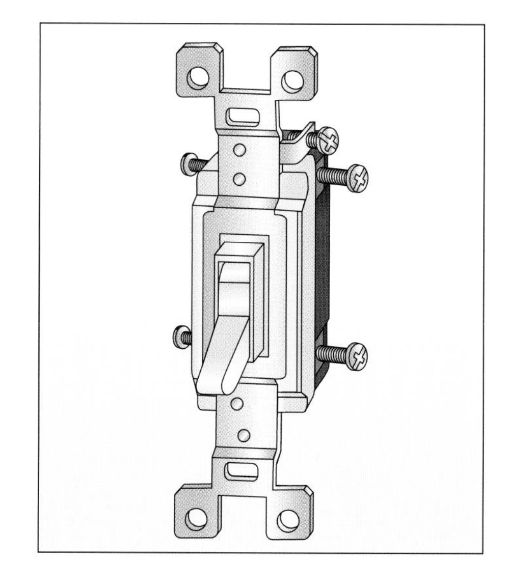

Four-Way Four-way switches aren't common—they're used to control a light fixture or receptacle from three or more locations. If your home has large rooms or long hallways, you may come across one. As with single-pole and three-way switches, a four-way switch is discernible by the number of screw terminals—it has four: two sets of "travelers." Four-way switches are always installed between a pair of three-way switches (*see page 100* for typical wiring).

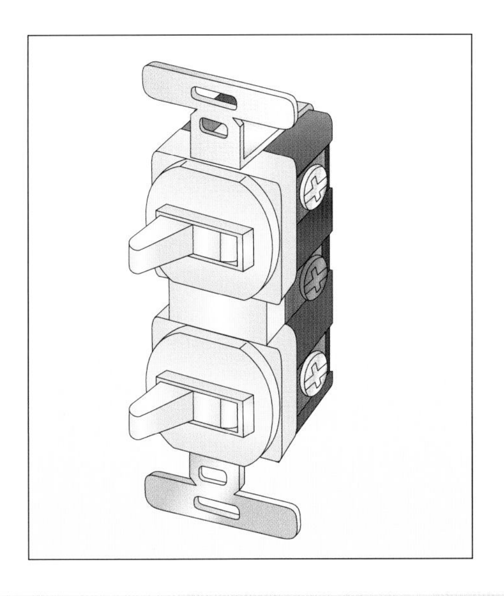

Double Switch A double switch controls two fixtures or receptacles from a single location. On most installations, both switches are powered by the same circuit—one wire (referred to as the feed wire) is connected to both halves of the switch; then two separate wires run power to the individual fixtures. For installations where there are two power wires, the metal connecting tab that joins the terminals is removed.

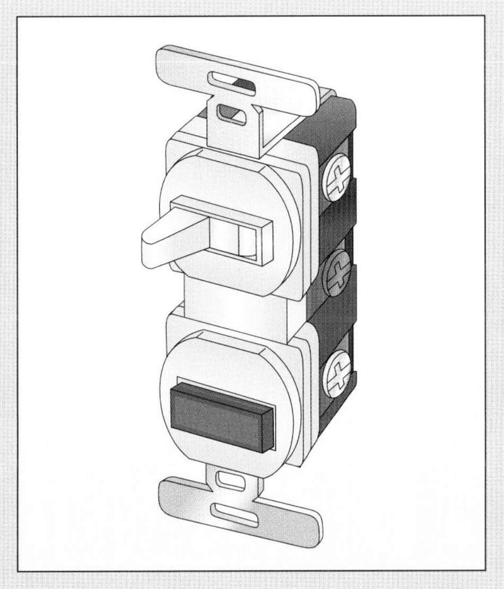

Pilot-Light Switch Pilot-light switches have a built-in indicator that lights when power flows through the switch and the fixture. These are especially useful for controlling devices that you can't readily see—like exhaust fans, garage lights, and so forth. Since this type of switch requires a neutral wire, it can't be used to replace a single-pole switch, where there are only two hot wires in the box.

SWITCH GRADES

As with any electrical part, there are two basic grades of switch available: residential and commercial. A residential-grade switch *(left)* is normally rated at 120 volts. They're inexpensive and are often shoddily made. A commercial-grade switch *(right),* costing about a dollar more, will serve you well for many years. These switches (rated up to 240 volts) have thicker bodies, corrosion-resistant contacts made from high-quality metal, heavy-duty nylon toggles, and larger, easier-to-use screw terminals. They're well worth the money.

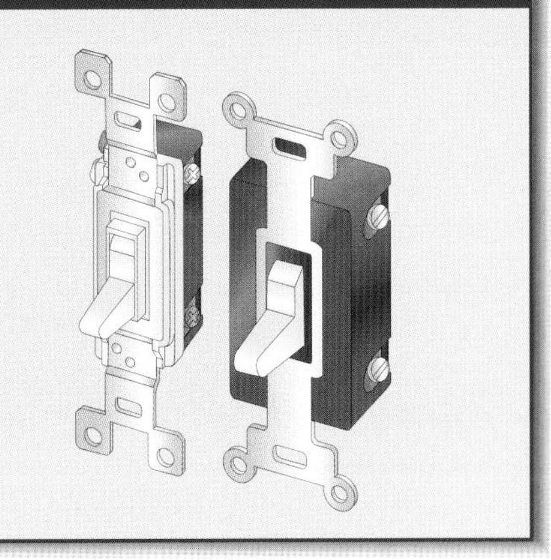

Specialty Switches

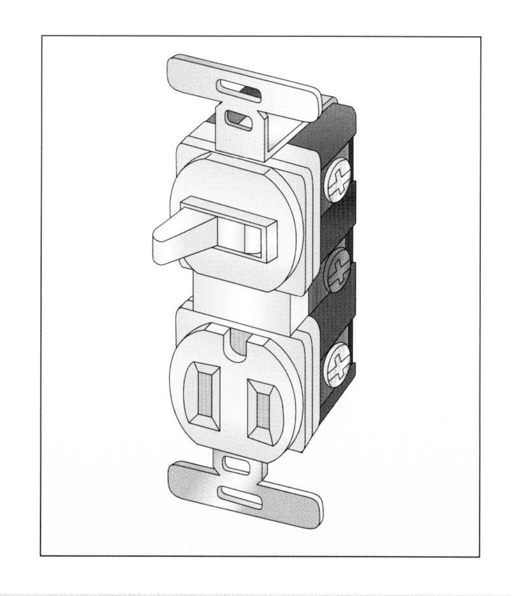

Switch/Receptacle Switch/receptacles can be used in two different situations. First, if you don't have enough receptacles in a room, you can add one by installing a switch/receptacle and wiring the receptacle so that it's always hot or ON. Or, you can wire the switch/receptacle so the receptacle is only hot or ON when the switch is ON; this is a handy way to remotely control the power to a plug-in device.

Time-Delay Time-delay switches allow you to turn a fixture or appliance on for a preset time. These can be electronic or manual (an internal spring is wound by turning the switch handle). After the allotted time that you selected has expired, the power is turned off. This type of switch is often used to control bathroom exhaust fans or heat lamps, and whirlpool bathtubs.

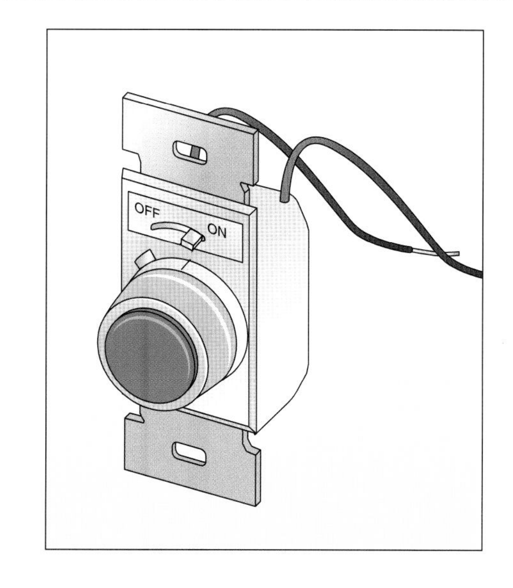

Timer A timer switch turns fixtures or appliances on and off at preset times. This type of switch is typically found in a living room, kitchen, or family room to turn lights on and off to simulate activity in a house. Timer switches can use either a time wheel or a digital clock to program the desired on/off cycles. Either type requires a neutral wire; therefore, a single-pole switch (which uses only two hot wires) cannot be replaced with a timer switch.

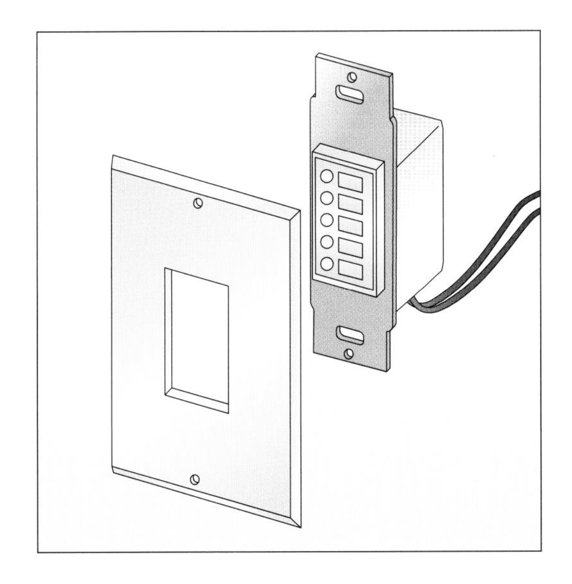

Motion Sensor A motion sensor switch employs the same technology used by security motion detectors: An infrared beam is projected out over an area to be protected. When motion is detected, the switch turns power on to a fixture or other device. After a preset time, it switches power off (if no further motion is detected). Motion-sensor switches don't need a neutral wire, so they can replace a single-pole switch.

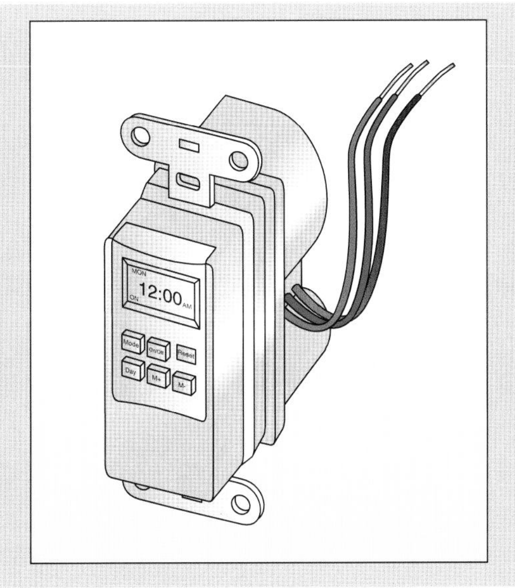

Programmable A programmable switch performs the same basic function as a timer switch but with one major difference: It can provide numerous on/off cycles each day to vary the lighting pattern. Varying the light usage in a random pattern better simulates normal use and is a proven crime deterrent. This type of switch does not require a neutral wire and is connected to your wiring with wire nuts.

Dimmer Dimmer switches let you vary the brightness of a lighting fixture. They're available in a wide variety of styles: slide action, lever type, and the most common—the rotary type (*shown here*). You can replace most standard single-pole switches with a dimmer switch; but manufacturers also produce three-way dimmers to replace standard three-way switches. All dimmer switches have wire leads instead of screw terminals—they're connected to the circuit with wire nuts.

Receptacles

Receptacles are manufactured to deliver an assortment of amperages at either high (240-volt) or low (120) voltage. Whenever you need to replace a receptacle, you should exchange it with an identical part. Check the back of the receptacle for both a voltage and current rating, or take it with you to the home center or electrical supply house.

Although many receptacles look similar, you can tell the difference between them by examining the configuration of the slots. These will indicate the voltage and current that the receptacle is capable of handling (*see the photos below and on page 41* for some typical configurations). If you have old-style, two-slot receptacles in your home, you can't just replace them with grounded receptacles; doing so without creating means for grounding makes the circuit unsafe.

120V, 15-amp – Ungrounded Two-slot receptacles, like the one *shown,* are common in older houses (pre-1960s) that don't have ground wires in the circuits. If wired correctly, the wide slot is neutral and the narrow slot is hot; this ensures that the polarity will be correct for any polarized device plugged into the receptacle. Don't replace a two-slot receptacle with a grounded receptacle unless there is some means of grounding available at the box.

120V, 15-amp – Grounded The 15-amp grounded receptacle *shown here* is one of the workhorses in a home. Here again, the wide slot is neutral and the narrow slot is hot. The big difference is the presence of a U-shaped grounding hole to accept the grounding prong of a plug. If wired correctly (*see page 93* for common receptacle wiring), this type of receptacle ensures both correct polarization and safe grounding of the circuit.

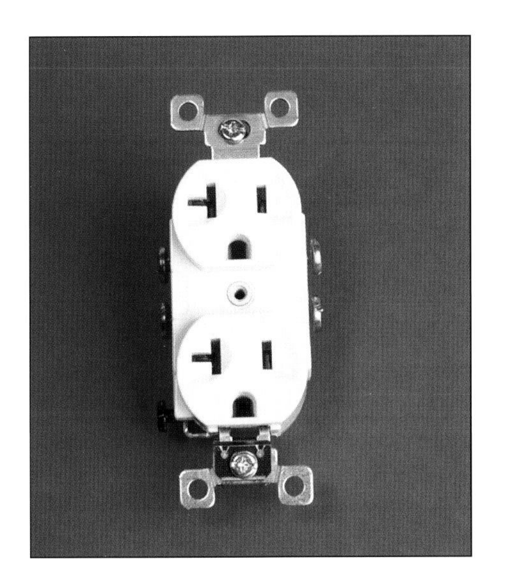

120V, 20-amp You can easily distinguish a 20-amp grounded receptacle from a 15-amp grounded receptacle by looking for horizontal slots on the neutral terminals. Although these T-shaped slots will accept a 20-amp plug, they will also accept a 15-amp plug. The reverse is not true, however; you can't insert a 20-amp plug into a 15-amp receptacle.

120V, 20-amp Single When you've got a critical device or appliance that you don't want possibly overloading a circuit, use a 20-amp single receptacle like the one *shown.* This type of receptacle is often used for window air conditioners, which would normally trip a circuit breaker or blow a fuse if another device on the circuit were to kick on.

240V, 30-amp Single The receptacle *shown here* is one of a couple of configurations available for supplying 30 amps at 240 volts. This type is commonly used for clothes dryers and some brands of large window air conditioners. If you're not comfortable handing 240 volts, call in a licensed electrician to replace or install one of these.

GFCI Receptacles

GFCI (ground-fault circuit interrupter) receptacles are safety devices designed to turn off power when a ground fault occurs. Basically, a ground fault occurs when current that normally flows through a circuit is shunted away from a circuit to flow where it shouldn't—like through a person. An electronic device inside the receptacle called a comparator constantly monitors both legs of a circuit—the hot side and the neutral or return side.

During normal operation (*see the drawing below*) the current flowing into a circuit or device should be the same as that flowing out of the device or circuit. If they're not the same, some of the current is flowing where it should not. This can occur when an appliance or device is faulty, or when an accident occurs, such as dropping a plugged-in hair dryer into a full wash basin.

If the receptacle is a GFCI-type, it will shut off power almost instantaneously. That's why code now requires GFCI outlets in the kitchen and bathrooms. Code also requires them to be installed in other locations such as in pool houses, garages, crawl spaces, outdoors, mobile homes, and even recreational vehicles.

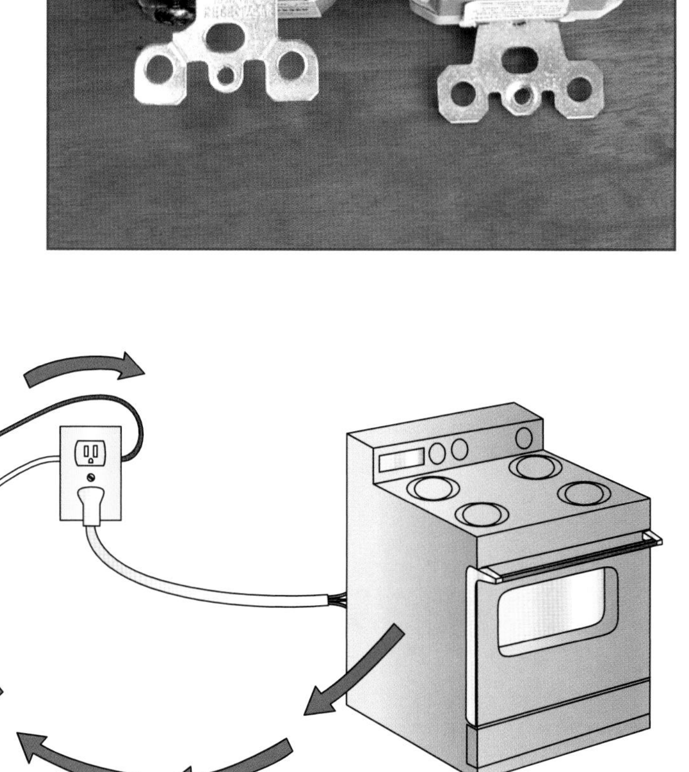

GFCI receptacles are capable of protecting both the devices plugged into them and all receptacles, switches, and light fixtures from the GFCI "forward" to the end of the circuit. They cannot protect devices between the GFCI and the main panel. (For more on GFCI wiring, *see page 96.*)

As with other safety devices in the home (such as smoke detectors), GFCI receptacles should be tested on a regular basis for proper operation. (*See page 85* for more on this.)

Cover Plates

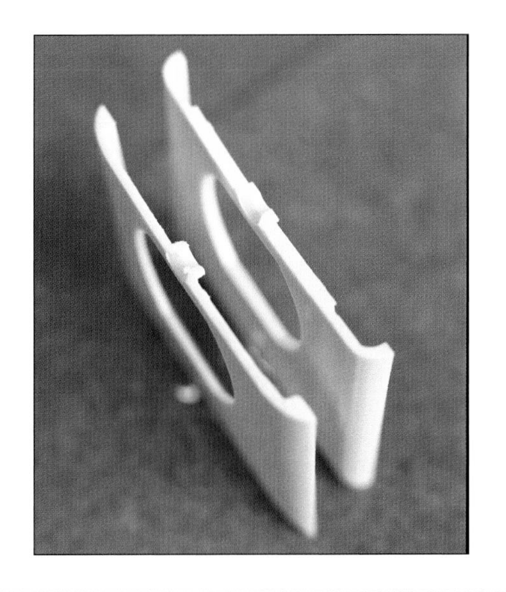

Grades What's the difference between "residential" and "commercial" grades—besides the cost? (Commercial plates typically cost 20 to 30 cents more than residential.) The difference can be seen clearly when you cut them in half. The thinner residential version (*front*) will easily crack when the cover plate screw is tightened. The beefier commercial plate (*back*) will stand up much better over time—it's well worth the extra quarter.

Outdoor and Utility Outdoor switches and receptacles need a cover plate that will protect them from the weather. This type of cover plate (*top center*) comes with a rubber gasket that fits between the cover plate and the box to form a watertight seal. Round-edged face plates for utility boxes are made to handle almost any situation. *Clockwise from right:* a single receptacle cover, a double box cover with a single receptacle opening, and a single switch plate.

Variations In addition to covering the electrical boxes, cover plates can add a decorative touch to a room. *Clockwise from left:* In stores and catalogs, you'll find everything from solid brass, vinyl-coated metal, fun plates for kids' rooms, and solid wood to the standard plastic plates in brown, almond, and white. Whichever type you buy, check the packaging to make sure it contains the necessary cover plate screws (you'll often find a package that's short a screw or two).

Connectors and Restraints

Wire Nuts Wire nuts are used to connect wires together quickly (*see page 53*) and are color-coded so you'll know how many wires you can safely splice. The most common are yellow and red. Yellow can splice three 12-gauge wires, or four 14-gauge wires; red can handle three 10-gauge, four 12-gauge, or five 14-gauge wires. Quality wire nuts contain a square-cut spring that cuts into and grips the wires as it's twisted into place.

Cable Staples and Wire Ties Cable staples (*left*) secure cable—typically non-metallic—to a framing member or wall. They all have built-in nails or staples for ease of installation. Just position the staple over the cable at the desired location, and hammer in the fastener. Wire ties (*right*) are often useful for neatly gathering cables or wires together. They are particularly handy in and around main service panels to group wires together.

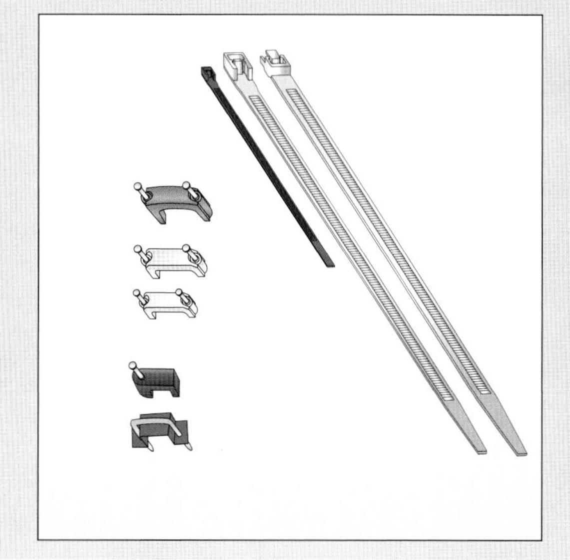

Conduit Straps Conduit straps are used to secure conduit to walls, ceilings, etc. "Saddle-style" clamps slip over the conduit and are fastened in place with nails or screws; you'll find them in both plastic (*top*) and metal (*middle*). Another type of conduit clamp is the hook-style, which can be hammered quickly into place (*bottom*). These can be driven into framing members or, with the use of plastic anchors, into masonry.

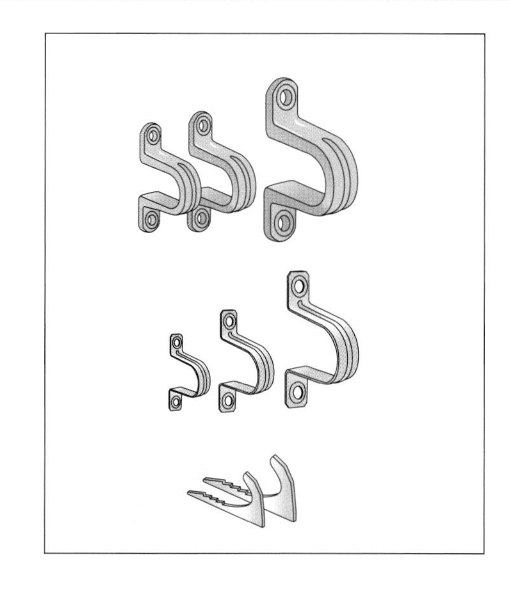

Fuses

Fuses are safety devices that "blow" when they detect a current overload. Although they can and do protect people, their primary responsibility is protecting the wiring in your home. Without them, too much current would continue to flow, the wires would overheat, and a fire could result. The two types of fuses you'll often come across in older homes are plug fuses and cartridge fuses. Plug fuses can be standard, time-delay, or tamper-proof. Standard fuses blow almost the instant current exceeds their designated limit. Time-delay fuses are designed to withstand a momentary surge of current, such as when an appliance starts.

The problem with both standard and time-delay fuses is they all have the same size threads. This allows someone to unscrew a 15-amp fuse and replace it with a 30-amp fuse. To prevent this, type-S, or tamper-proof, fuses and adapters were developed (*see below*). They have smaller plastic threads and are designed to fit only the correct-current–rated adapter.

Regular and Time-Delay Plug fuses are rated for 120-volt use only and are available in ½, 1, 2, 3, 5, 6, 8, 10, 15, 20, 25, and 30 amps. Fuses 15 amps and lower have a hex-shaped window opening; 20-, 25-, and 30-amp fuses have a round opening. Fuses with tops visible, *from left to right:* a 3-amp standard fuse, a 20-amp fuse with built-in circuit breaker, and a 30-amp time-delay fuse. A tamper-proof or type-S fuse and adapter are *shown in the inset.*

Cartridge Fuses Cartridge fuses are used to protect 240-volt circuits. Ferrule-type cartridge fuses (*shown here*) protect circuits up to 60 amps. Knife-blade cartridge fuses, identified by metal blades at each end of the fuse, can handle circuits above 60 amps, up to a high range of around 600 amps. Always remove and replace a cartridge fuse using a fuse puller (*see page 24*)—and make sure you're using an exact replacement fuse.

Circuit Breakers

Circuit breakers have replaced fuses as the primary means of providing over-current protection in modern homes. The big advantage they provide over a standard fuse is that they can be reset (*see page 69*) once they've "tripped"—standard fuses must be replaced. A circuit breaker contains a bimetal strip that heats up when too much current passes through it and causes the breaker to "trip," shutting off the power. (Some breakers have a built-in electromagnet that helps the breaker trip even faster when a short is detected.)

Circuit breakers are rated to handle 120-volt, 240-volt, or both 120-volt and 240-volt circuits and to provide current protection ranging from 15 to 200 amps. In addition to monitoring current, circuit breakers also are sensitive to heat. If there's too much ambient heat in the panel, a breaker (or breakers) can trip.

Single-Pole Single-pole circuit breakers control current for loads that use only one leg of the 240 volts available in a breaker panel. Standard single-pole breakers are rated to handle 15, 20, 25, 30, 40, or 50 amps. They come in different configurations to handle almost any situation, the most common being full (*left*) and dual (*right*). Some manufacturers also make half-sized breakers (½"-wide) for full panels; this way you can fit two half-sized breakers in one slot.

Double-Pole Double-pole circuit breakers control the current on loads that use both legs of the 240 volts available in a panel. They're rated from 15 amps all the way up to 200 amps. Thirty-amp breakers are often used for clothes dryers and water heaters. Breakers rated 40 (*left*) to 50 amps typically protect electric ranges; and breakers above 60 amps, like the 100-amp breaker *shown at right,* usually protect furnaces and other large-current–demanding devices.

Service Panels and Switch Boxes

Every home has a service panel. It accepts the cable coming from the power company via the service entrance (*see page 10*) and distributes power throughout the home by way of branch circuits. Each of these branch circuits is hooked up to a separate breaker. (Older service panels like the one shown on *page 12* use plug-in and cartridge fuses in lieu of breakers.)

Installing a service panel is a job best left to a licensed electrician. There are numerous code requirements, and power will have to be temporarily disconnected. An improperly installed service panel is dangerous and can cause shocks or even a fire.

Subpanels or switch boxes (also referred to as "lugs-only" panels, as they have no main breaker) are used to protect branch circuits going out of the main service panel (*see below*).

Service Panels Modern service panels, like the empty one *shown here,* are designed to handle anywhere from 100 up to 800 amps, with 200-amp panels being the most common in new homes. Each panel is designed to hold a specified number of circuit breakers. A main service panel will have a main breaker (also referred to as the main disconnect); a hot bus, which runs down the center of the panel; and a neutral/grounding bus located on the side opposite the hot bus.

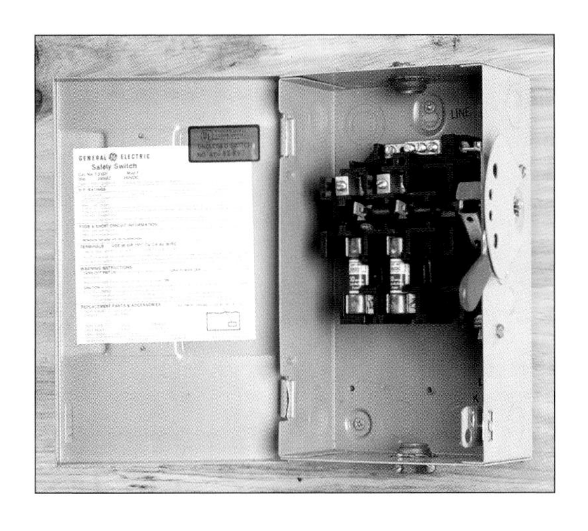

Switch Boxes Switch boxes or subpanels are used where multiple branches need to run a distance from the main panel—such as a new addition to the home. The advantage here is that you only need to run a single cable to the subpanel and then branch out, instead of running multiple cables—and if the distance is great, this can save a lot of money. Although the 60-amp cartridge fuse version *shown here* is quite functional, a breaker panel would be more convenient.

Chapter 3

Working with Wire, Cable, and Conduit

The key elements to any electrical project are the wire, cable, and conduit. These are the veins and arteries of the system that directs power to the various parts. And just like the human equivalent, a nick, break, or crimp in a wire or cable can create a dangerous situation. That's why it's so important to take the time to learn how to work with these materials from the beginning.

An inattentive job of stripping away insulation or cable sheathing can nick a wire. If the wire is then bent around a screw terminal and compressed via the screw head, the nick can degrade and cause the wire to break eventually. A loose, energized wire in an electrical box can create a short, a ground fault, or even a spark resulting in a life-threatening fire.

To work safely with wire, cable, and conduit, it's imperative that you have a solid understanding of the differences and applications of the various materials. Chapter 2 covers this in depth, and I urge you to review any section that applies to an upcoming electrical project.

In this chapter I'll begin by explaining how to cut and strip wire and cable (*pages 49–53*). Then we'll look at ways of joining wire together to make sound, safe electrical connections (*page 53*). Next, I'll go over preparing wires for connecting to both screw and push-in terminals on receptacles, switches, and fixtures (*pages 54 and 55*).

The second half of this chapter delves into working with cable: connecting cables to boxes to ensure a safe transition into the box, thereby eliminating any chance of harming the sheathing (*pages 56–57*); working with flexible armor-clad cable (*pages 58–59*); and cutting and bending metal conduit (*pages 60–63*); and on *page 64,* how to pull wire through conduit (there's more to it than just muscles and a strong back).

If knocking holes in your walls makes you nervous, I've included a couple of pages on surface-mount wiring—no holes necessary—just attach the metal or plastic "raceway" directly to your walls (*pages 66–67*).

Stripping Wire

There are two basic ways to strip wire: with a pocket- or utility knife or with a pair of wire strippers. If you've got only a few wires to strip, a pocketknife will work just fine. However, if the project you're working on requires a lot of electrical connections, it's worth the expense to purchase a pair of wire strippers. There are two main types: One type (shown in *the lower photo below*) has a series of notches that accept the different gauges of wire; the other type has a screw that slides in a slot and can be adjusted to strip the gauge you're working with. The advantage of this type is that once the stripper is set, you don't have to locate the correct notch every time. The disadvantage is that you'll have to adjust the screw position if you're working with different gauges of wire.

With a Pocketknife I've got an electrician friend who can strip a piece of wire with a pocketknife in one fluid motion—faster than I can with a wire stripper—he spins the knife around the wire and pulls as he completes the cut. It takes a bit of practice before you can do this without nicking the wire—something you don't want to do, since a nick in the wire can lead to premature breakage when the wire is bent.

With a Wire Stripper Stripping wire with a wire stripper is easy. Start by matching the gauge of the wire you're working with to the corresponding notch in the stripper. Insert the wire and close the jaws firmly on the wire. Rotate the stripper roughly 180 degrees while holding the wire to prevent it from twisting. This cuts the insulation fully so you can then pull the wire stripper to slide the cut insulation off the wire.

Cutting Wire and Cable

Diagonal Cutters I often use insulated-grip diagonal cutters (commonly called "dikes" in the trade) to cut individual wires in the 10- to 22-gauge range (*see the wire color/gauge charts on pages 26 and 27*). The tapered point allows me to reach into places that linesman pliers (*below*) can't get into. I don't advise cutting anything heavier than 10-gauge with these: If you do, you run the risk of breaking the cutter or "springing" the jaws.

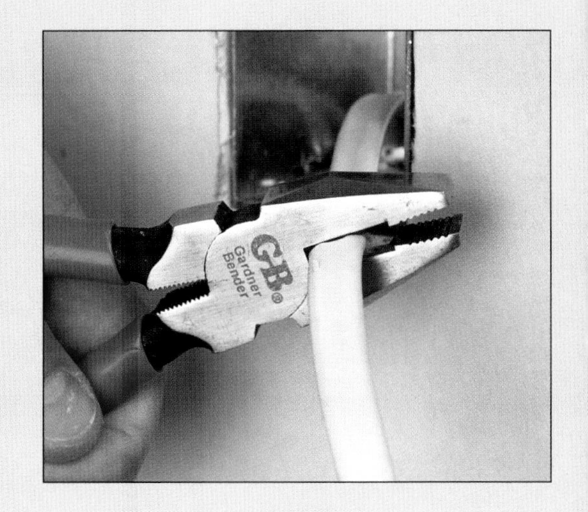

Linesman Pliers The number one tool that most electricians reach for is the larger, beefier linesman pliers. This tool is capable of handling larger-gauge wires and even thick two- and three-wire non-metallic and underground cable. When cutting heavy-gauge wire or cable, position the wire as far down into the jaws as possible to concentrate the cutting action. Linesman pliers also feature square, serrated tips that allow you to grip, pull, and bend wire or cable.

Nippers Another cutting tool I keep handy for electrical projects is a pair of nippers. They're great for cutting wire staples and are absolutely invaluable for reaching into electrical boxes and cutting wires where clearance is a problem. They're also great for flush-cutting protruding wires and for general-purpose trimming.

Preparing NM Cable

Non-metallic (NM) cable is by far the most popular choice for home wiring projects, and for good reason: It's so easy to work with. Whether you're adding a receptacle (*page 116*) or installing NM cable in new construction (*pages 122–123*), the first step is to access the wall studs. Then drill holes in studs to route the cable to its intended destination. Once the cable has been inserted and secured into an acceptable electrical box (*pages 87–92*), the ends of the cable can be prepared for connections to a switch, a receptacle, or a light fixture. This involves stripping off the outer sheathing to expose the wires, cutting the wires to length, and then stripping the wires (*page 49*). To remove the outer sheathing, you'll need to cut it down its length, peel it back, and then cut it away.

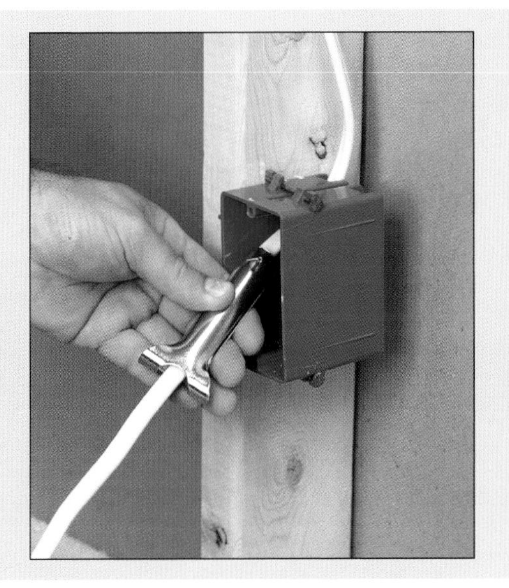

1 Position the cable ripper Although you can use a utility knife to cut into the outer sheathing, you run the risk of cutting into the insulation of the wires and possibly nicking the wire. A safer solution is to use a nifty tool called a cable ripper. It costs only a few dollars and was designed just for this purpose. To use the cable ripper, position it about 8 inches from the end of the cable and squeeze it so the cutting point penetrates into the sheathing.

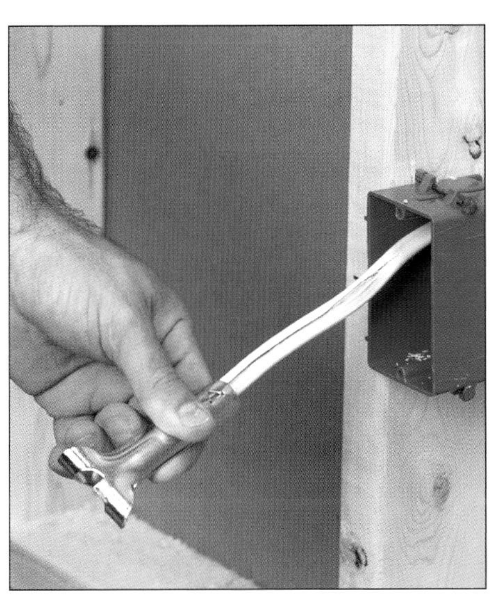

2 Pull to strip Now grip the cable ripper firmly in one hand while securely holding the cable in the other. Pull the cable ripper toward the cut end as you continue to squeeze the ends of the cable ripper together. If it doesn't rip cleanly, the cutting point may be dull. A few strokes of a small, flat mill file will quickly bring the point back to a keen edge.

3 Peel back the sheathing After you've split the cable with the cable ripper, you can peel back the outer sheathing to expose the wires. In most non-metallic cable, you'll also often find paper either wrapping just the ground wire or wrapping all of the individual wires inside the cable; peel this paper wrapping off as well to expose the wires.

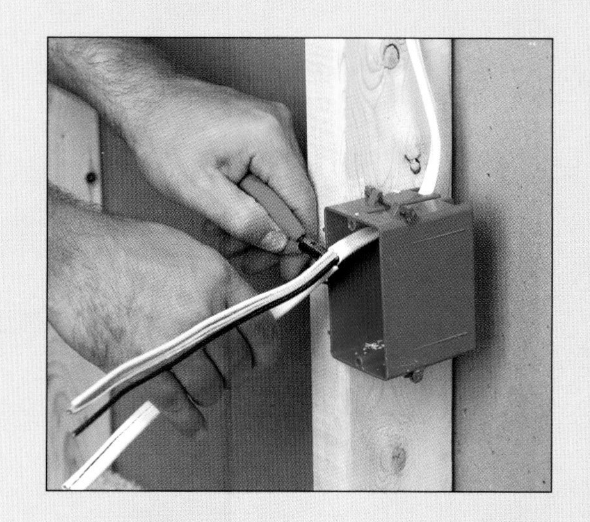

4 Cut off excess sheathing Now that you've peeled back the sheathing, you can cut off the excess. I like to use a small pair of diagonal cutters for this, since it allows me to cleanly trim the sheathing near the base of the cut. Linesman pliers or a utility knife will work fine, too; just be sure not to nick the wires. Trim away any paper wrapping as well.

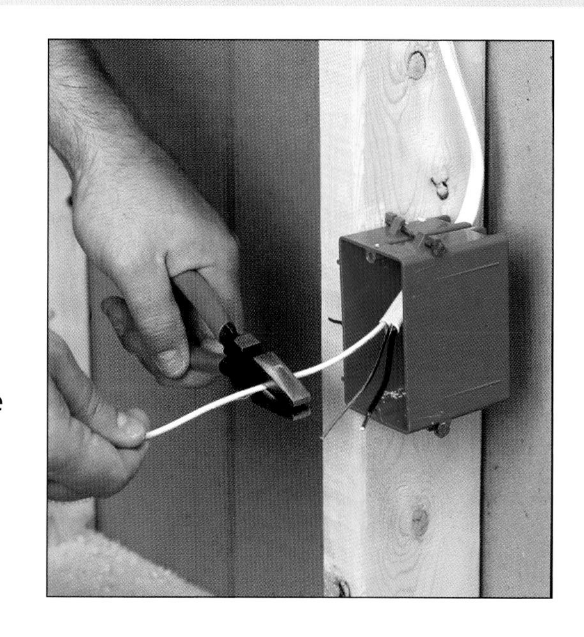

5 Cut wires to length and strip The final step before making electrical connections is to cut the wire to length and strip it. Trim the wires to a length of around 4 inches. This leaves plenty of clearance to wrap around most switches and receptacles; it also provides a margin of error in case you nick a wire and need to trim off the end and strip the wire again (*see page 49* for more on stripping wire).

Joining Wire

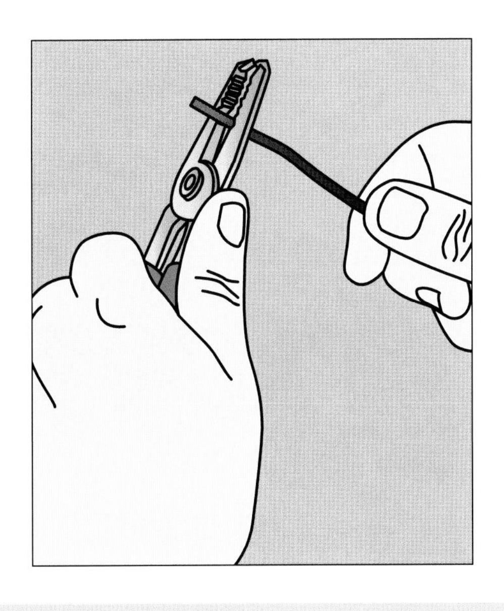

1 Strip to same length To join together two or more pieces of wire with a wire nut, begin by stripping the wires (*see page 44* for help in choosing the correct size wire nut). In most cases, strip off about ¾" of each wire to be connected, using a wire stripper (*as shown*) or a pocketknife (*see page 49*).

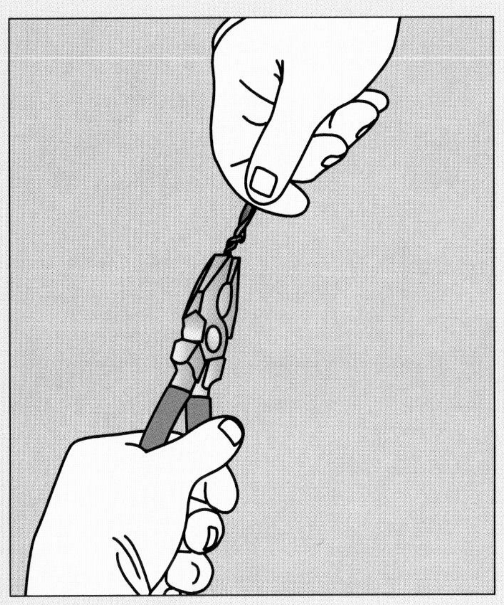

2 Twist with linesman pliers Although you can use a wire nut without twisting the wires together, I like to twist them together in a "pigtail" splice as added insurance—especially if the connection will be exposed to vibration or repeated cycles of hot and cold (which can work the wires loose over time). Use a pair of linesman pliers to grip the tips of the wires. Then, while holding the wires firmly in your other hand, twist the linesman pliers.

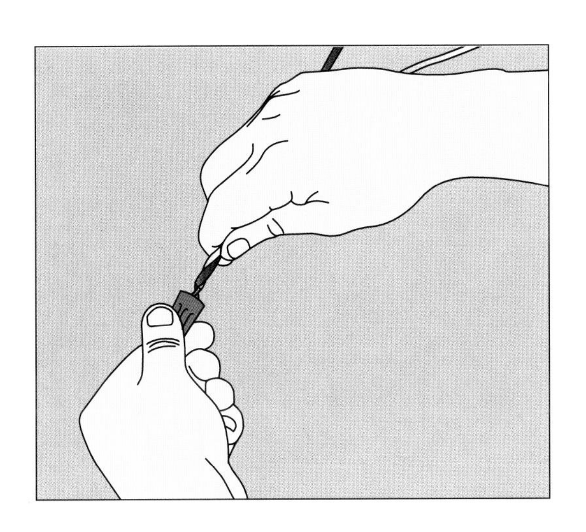

3 Screw on the wire nut If necessary, trim the pigtail so that when the wire nut is screwed on, no bare copper is exposed. Turn the wire nut clockwise to tighten it, counterclockwise to loosen. Twist the nut until you meet firm resistance. Then gently pull each of the wires to make sure that the connection is solid and that there's absolutely no chance of the wires coming loose.

Preparing Wires for Screw Terminals

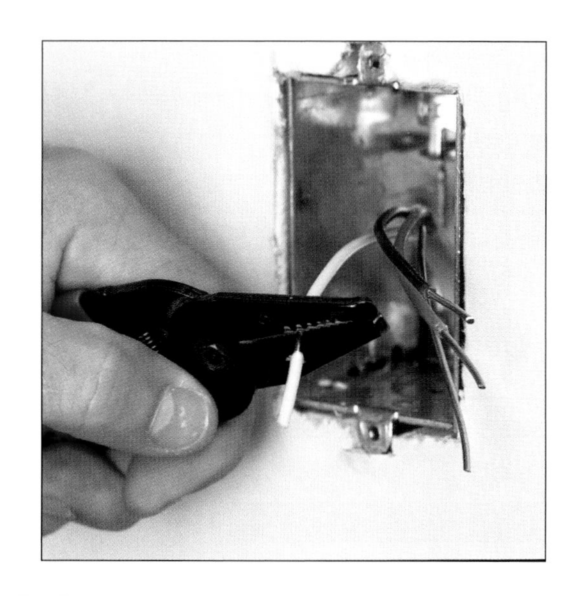

1 **Strip wire** To prepare wires to attach to screw terminals, strip about 3/4" off of each wire with a wire stripper or a pocketknife. Since you'll be bending this wire into a loop in the next step, it's particularly important that you not nick the wires. If you accidentally do, though, trim off the end of the wire at the nick and restrip the wire.

2 **Form loop with pliers** The next step is to form a C-shaped loop on the end of each wire to wrap around the screws. A pair of needle-nose pliers works best for this. If you grip the end of the wire so it's about 1/4" into the jaws and then twist, you'll form a perfect loop for the screw terminal. Here again, it's critical that you don't nick or scratch the wire.

3 **Hook onto terminal** Hook each wire around the appropriate screw terminal so that it forms a clockwise loop; this way, as you tighten the screw, it will pull the loop tighter around the terminal instead of forcing it open. Tighten the screw firmly, and make sure that no insulation is captured under the head of the screw. Never place two wires under a single screw terminal; instead, run a single wire to a pigtail splice (*see page 53*).

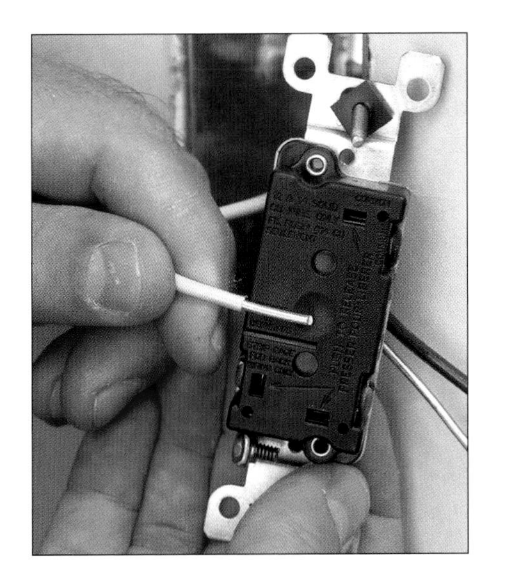

Preparing Wires for Push-In Terminals

1 **Use gauge and strip wire** In addition to screw terminals, some receptacles and switches are designed with terminals that accept a wire that's pushed into an opening in the back of the device. A spring inside grips the wire and makes a quick electrical connection. There's a gauge on the back that indicates how much wire to strip. There's also a corresponding opening where each wire enters where you can release the wire (*see Step 3 below*).

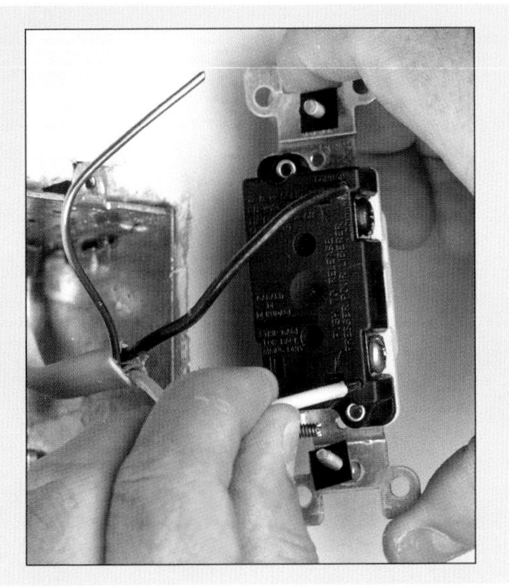

2 **Push the wire in** Once the wire has been stripped, insert each of the stripped ends into the appropriate opening. Push the wire firmly into the opening so that no bare copper is exposed. Then gently tug on each wire to make sure the spring is gripping the wire securely. If copper is exposed, or if the wire slips, release the wire (*see Step 3*) and try again.

3 **How to release** If you need to remove a wire that's been inserted into a push-in terminal, locate the square or rectangular hole adjacent to the wire. Insert a pick (*as shown*), a nail, or a small flathead screwdriver into the opening and press down. As you maintain pressure with the pick or screwdriver, pull up gently on the wire to remove it.

Connecting Cables to Boxes

Connecting a cable to an electrical box may seem like an unimportant detail, but it's not. In fact, an improperly connected cable is one of the major causes of electrical problems. The reason for this has to do with the opening in the box that the cable passes through. On both plastic and metal boxes, these openings usually have jagged edges. If the cable isn't securely fastened to the box, it's possible for it to move due to vibration, house settling, or exposure to repeated hot and cold cycles. This movement can eventually abrade the sheathing and insulation, resulting in wires touching metal or other wires. You can prevent this by making sure the cable is securely fastened to the box. This can be achieved with quick clamps or metal saddle clamps inside the box, staples outside the box, or connectors that screw onto the box.

Plastic Box with Metal Clamps Many plastic boxes (especially those used for new construction) have internal metal saddle clamps to secure cables. To use one of these, loosen the clamp until you can slip the cable through the opening and under the clamp. Pull the cable through until you end up with 8" to 10" to work with. Then tighten the saddle clamp to grip the cable firmly. Give the cable a gentle tug to make sure there's no play.

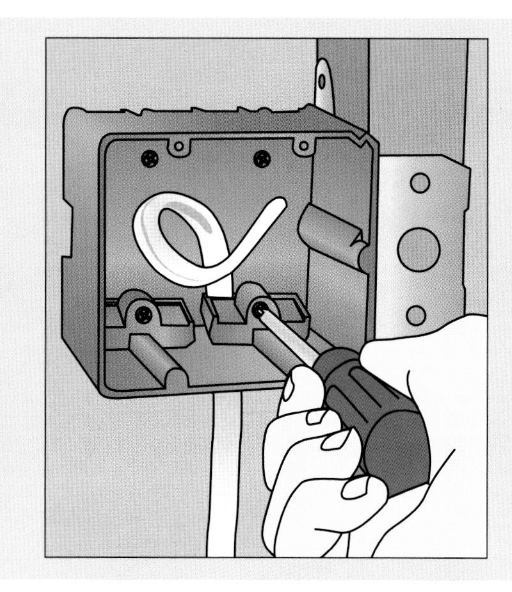

Plastic Boxes with No Clamps or Staples On some plastic boxes, there are no built-in cable clamps. This type of box should be used only for new construction, where you have full access to the wall studs. This way you can insert the cable in the opening in the box and then secure it to the wall stud with insulated cable staples (*see page 44*). Position a couple of staples near the box to hold the cable in position.

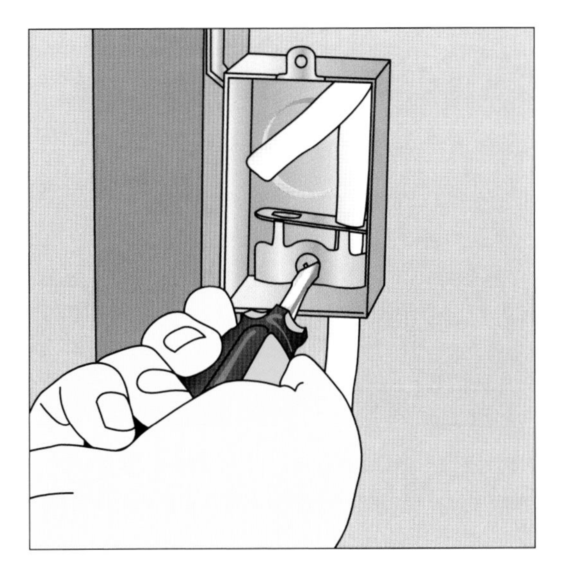

Metal Box with Saddle Clamp A wide variety of metal boxes also have built-in saddle clamps. The procedure for using them is similar to that for the plastic boxes shown on *page 56.* The only thing to be aware of here is that it's easy to overtighten the clamp and damage the wires since you're screwing into metal threads (unlike the threads on a plastic box, which can strip if overtightened). Remember, the goal here is to keep the cable from moving over time.

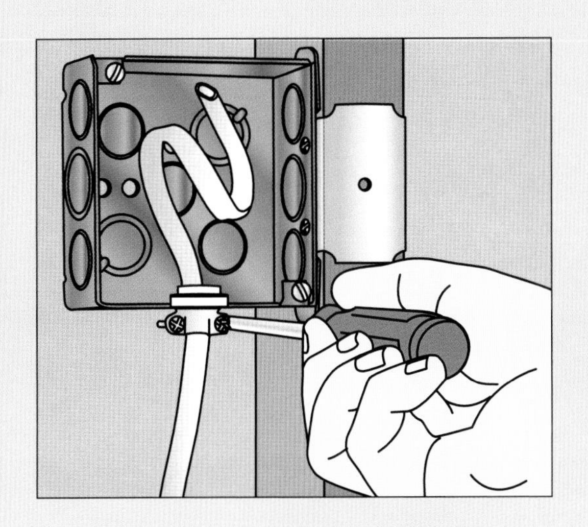

Metal Box with Clamp Connectors The cable connector that I prefer is the type that screws into an opening in a metal box. This style of connector accepts a wide variety of cable (non-metallic and armored) and conduit (*see page 30* for more on connector types). These connectors eliminate the abrasion problem since the wire passes safely through it as it passes into the box. The connector attaches to the box with a locknut; the cable is secured to the connector with a pair of screws.

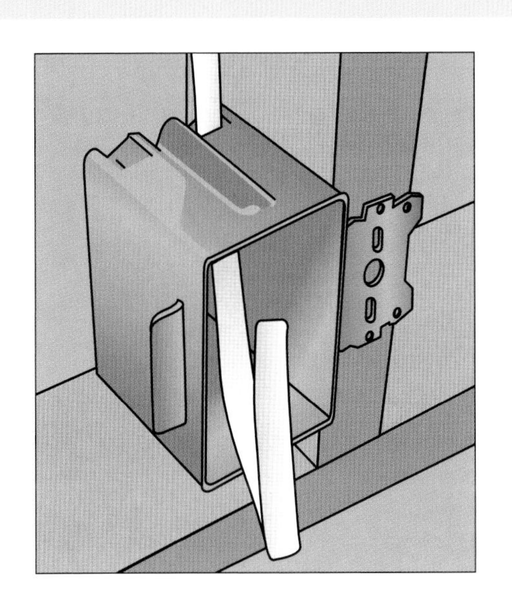

Quick Clamp A relative newcomer on the plastic box market, this type of box features built-in spring clamps to hold cables securely in place. A cable is inserted though the opening and pulled through to the desired length. The edge of the plastic spring digs into the sheathing to capture it. Caution: Don't pull the cable the reverse direction; the spring is likely to cut into the sheathing. If you need to pull some cable back, insert a flathead screwdriver between the spring and cable before pulling.

Working with Armor-Clad Cable

Armor-clad cable, or BX (a trade name it's commonly referred to as), is a flexible metal cable that can carry two, three, or four conductors wrapped with paper. There's also a 16-gauge aluminum "bonding" wire that runs the length of the cable; it's attached to the armor along its length to ensure that the flexible links are properly grounded. It should never be used as the grounding wire. Since armor-clad cable is restricted in many applications, check with your local building inspector before installing it in your house. I've most often seen it used in industrial applications where power needs to be connected to a unit that has to be periodically moved—like a kitchen appliance that needs regular cleaning. In some locales, BX can be installed just like non-metallic cable (*see pages 120–123*). The only real tricks to working with armor-clad cable are cutting it and then terminating the ends properly (*see below*).

1 Cut the sheathing Start by clamping the cable securely to a sawhorse or workbench, and then mark the cable about 8" in from the end. Then with a hacksaw held at a right angle to the spirals in the cable, cut into the sheathing at the mark. Stop as soon as you've passed through the sheathing. Now twist the armor until it snaps free. Cut off any paper wrapping the wires, and the bonding wire, with a pair of diagonal cutters or linesman pliers.

2 Thread through the hole Once you've prepared the end, you can thread the cable though the holes in the wall studs and framing members that you've drilled. Since armor-clad cable is heavier-duty and stiffer than non-metallic cable, it doesn't turn corners as easily—it requires a much larger turning radius. To test, simply form a piece of the cable into a loop to determine how tight a corner you can turn.

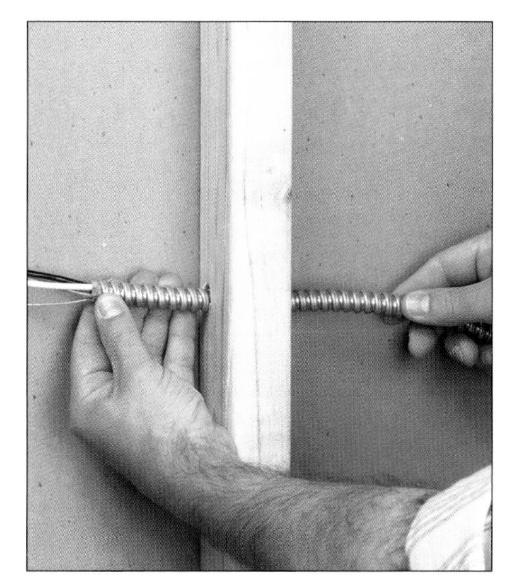

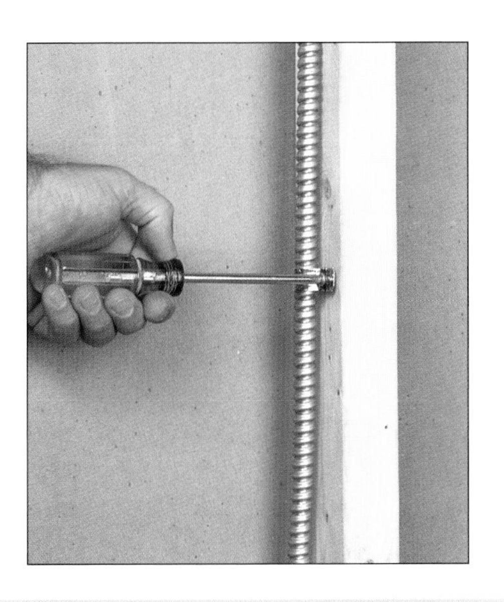

3 **Secure cable** Because of its substantial weight, armor-clad cable needs to be securely fastened to wall studs and framing members to prevent it from pulling on boxes and fixtures. Use screw-in or nail-in cable straps (*see page 44*) to fasten the cable in place every couple of feet and especially near the curves. Give the cable a gentle tug to make sure it's held fast.

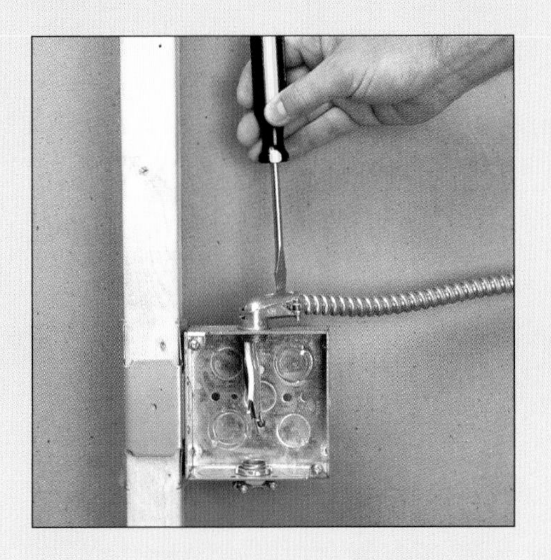

4 **Attach cable to connector** Pick the armor-clad cable connector to fit how your cable is entering the electrical box. The one *shown here* is a two-piece unit designed to enter the box at a right angle. Remove the cover piece (it's held in place with two screws), and insert the cable so the cut end of the cable will be firmly held inside the connector (this establishes the ground). Then slip the cover back in place and tighten the screws.

5 **Tighten the locknut** Finally, to secure the cable connector to the box, slip the locknut over the wires and thread it onto the end of the cable connector. To lock it in place, place the tip of a flathead screwdriver against one of the tabs on the locknut and give it a firm rap with your hand or a hammer; clockwise tightens, counterclock-wise loosens.

Working with Conduit

I'm not sure what it is that scares many people away from working with conduit. I really enjoy it; it's kind of like working with Tinkertoys. The only difference is you can cut the conduit to any length you want—and you get to bend it (*see page 62*). Now don't get me wrong—there is an art to bending conduit—but a simple job without a lot of bends is something that almost anyone is capable of.

All you'll need are a couple of special tools. First, you'll need a hacksaw or a tubing cutter to cut the conduit; you'll get a cleaner, more accurate cut with the tubing cutter. Second, if bends are involved, you'll need a conduit bender. You can find this at any home center. As with any project, it's best to have all your supplies (conduit, electrical boxes, etc.) on hand before you start the job. That way, you'll be able to assemble as you go.

1 Measure and cut With a tape measure, measure from box to box or from a box to a connector. Then subtract the length of the offsets or connectors you'll be using. Transfer this measurement to the conduit, and make a mark. Place the cutting wheel of the tubing cutter on the mark, and tighten the knob. Then spin the tubing cutter around the pipe and tighten the knob as you go. Continue tightening and spinning until you cut though the conduit.

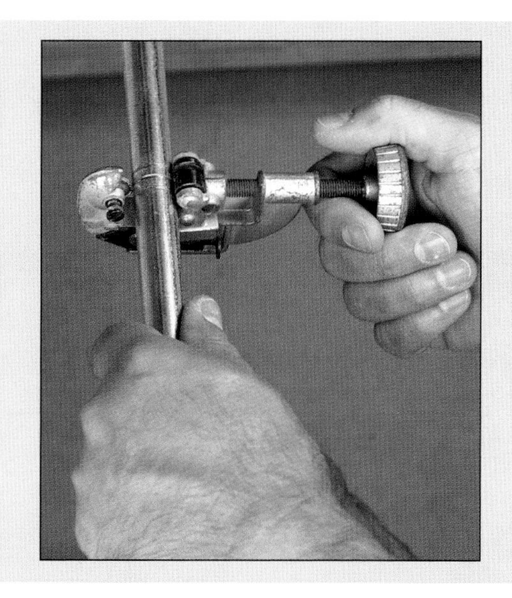

2 Ream the ends The next step is to ream out the ends of the conduit to remove any sharp edges that could cut into sheathing or insulation. Most tubing cutters have a built-in reamer (or you can purchase a T-handle–type reamer if you've got a lot of conduit to run). Insert the reamer in the end of the conduit and, while maintaining firm pressure, spin the tool. Continue until you can't feel any burrs or rough edges with your fingertip.

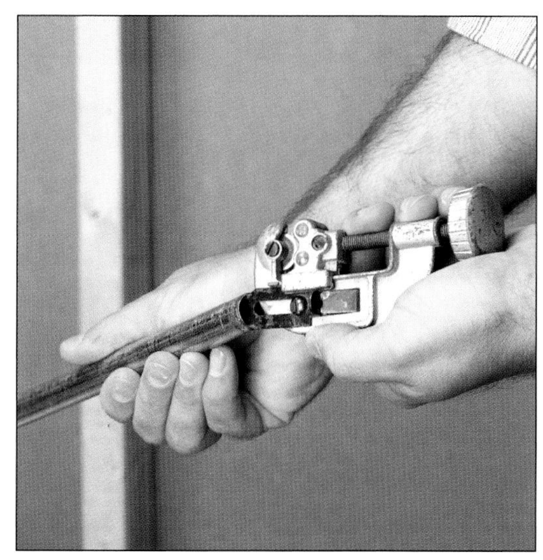

3 Install boxes After you've reamed out the conduit, test-fit the pieces together (including the electrical boxes) and hold them in the position in which they'll be mounted. Make a pair of marks where each box will be attached to the wall. If you're mounting the box to a wall stud, simply screw it in place with panhead screws. On a masonry wall, you'll need to drill holes for concrete screws or anchors before fastening it in place.

4 Add offset if necessary In most cases where conduit comes into a box, you'll want to add an offset. This is just a connector that has a slight jog in it and serves as a transition between the box and flush-mounted conduit. Insert the offset in the desired hole in the box, and thread on the locknut. Use the tip of a flathead screwdriver to lock the nut firmly in place.

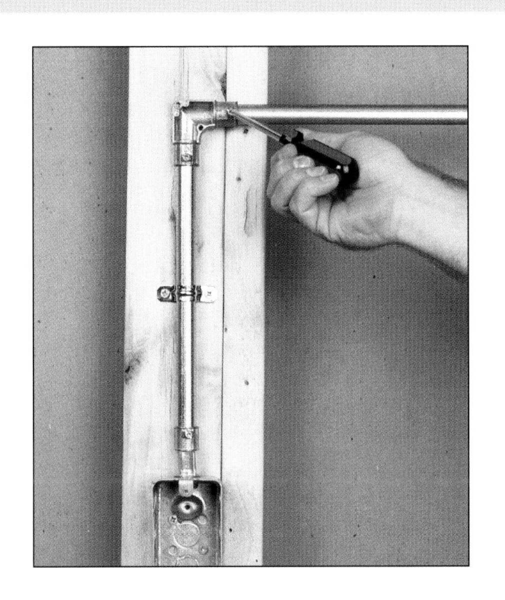

5 Install elbows For instances where you need to turn a corner, you can bend the conduit (*see page 62*), use prebent corners, or install an elbow. Elbows have a removable cover that provides access to the wire for pulling it through the conduit. Insert the conduit in both ends of the elbow, and tighten the setscrews to lock them in place. Remove the cover when you go to pull wire (*see pages 64–65* for step-by-step instructions).

Bending Conduit

Without a doubt, bending conduit is an art. I had the pleasure of working with a U.S. Navy installation crew when I was in the service. They started with an empty room and in a matter of weeks, left behind a functioning communications site. I was most impressed with an older fellow whose job was installing the conduit. I watched him one day (under the pretense of helping) as he installed a set of ¾" conduit. Each piece came out of a bay, made a 90-degree turn and then two more turns, and finally went off in another direction. As he worked, I was struck by a couple of things. First, his level of confidence was extraordinary: He just looked up and bent the conduit—no measurements. And each concentric bend nestled up next to the prior piece in a perfect fit. Truly amazing. The bottom line here is that bending conduit requires practice—make sure you buy plenty of extra lengths for any project you're planning. (Just imagine how many twisted lengths of conduit this fellow discarded in his early days.)

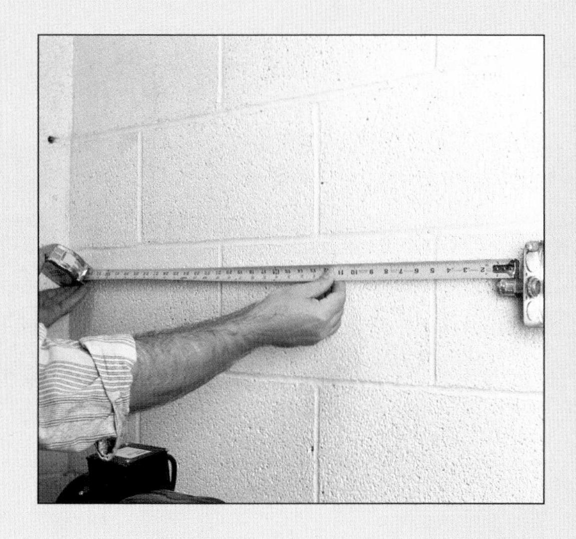

1 Measure The first step to bending conduit around a corner is to take a couple of measurements. With a tape measure, measure out from the corner to each box. Then subtract the length of any offsets or connectors you're planning on using. You'll also need to subtract the take-up (the amount that will be gained by the bend) from this length (5" for ½" conduit, 6" for ¾" conduit, and 8" for 1" conduit). Mark and cut the conduit to this length.

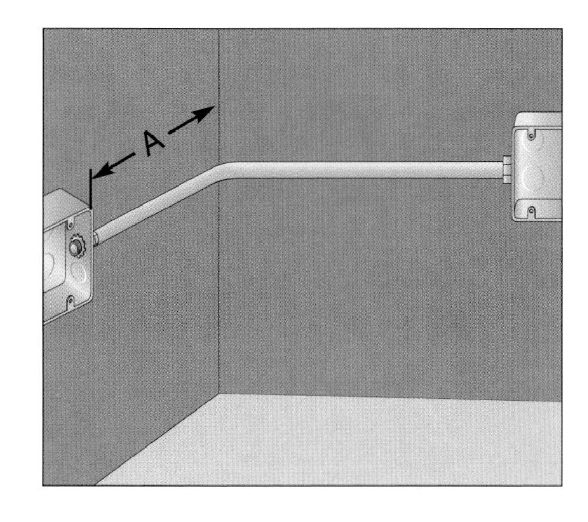

2 Position the conduit bender This is the tricky part—positioning the bender on the conduit in the proper place. Here's how to do it. Make a mark on the conduit equal to the distance "A" *shown.* Then subtract the take-up from this and make a mark. This is the mark you'll want to line up with the alignment mark on the conduit bender.

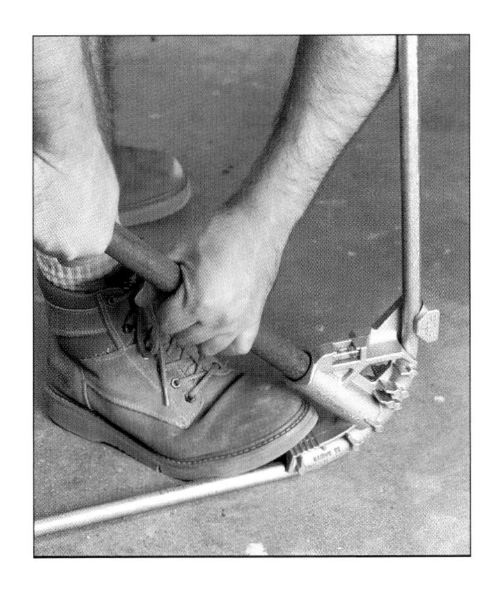

3 Make bend with conduit bender Place one foot on the "pedal" of the bender and with firm, gentle pressure, slowly pull the handle toward you. Gentle pressure is essential here, as it's very easy to put a crimp in the conduit—and installing crimped conduit is forbidden by code. Most benders have a built-in bubble level to help show you when you've achieved a 90-degree bend.

FORMING AN OFFSET

Although forming a transition or offset into an electrical box requires some practice, it's very satisfying when you get it right.

There are two tricks for this that may be of help. First, since the offset must be aligned with other bends, draw a reference line along its length as a sight aid. Second, whenever possible, make the offset first and then make any other bends. Or, if no other bends are involved, bend the offset, then cut the conduit to length.

To create an offset, start by making a 15-degree bend *(below left).* Then raise the conduit up off the floor by placing it on a scrap of 2×4. Now move the bender up a few inches and bend it in the opposite direction until the first bend is parallel with the floor (below *right*).

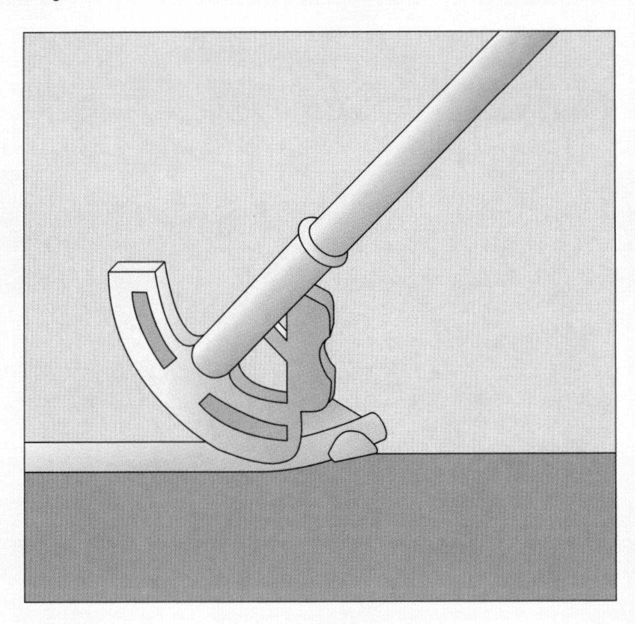

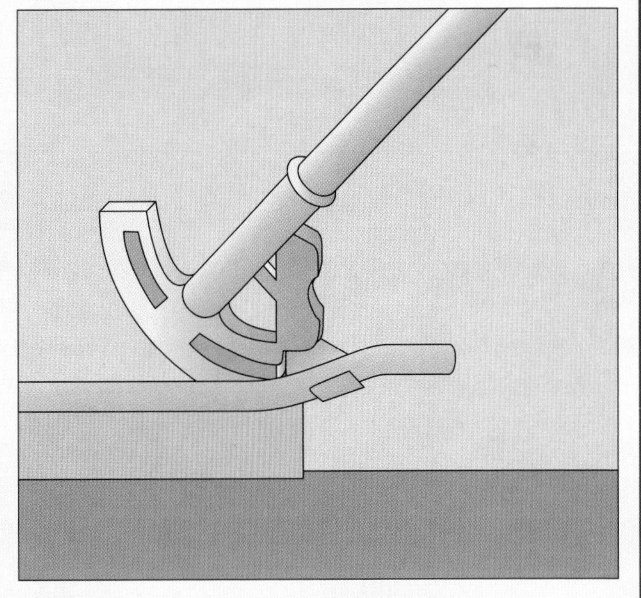

Pulling Wire through Conduit

Pulling wire through conduit can be hard work—especially if there are a lot of bends, if the run is long, or if you're running the maximum number of allowable wires through the conduit (*see page 30*). Although the electrical code's rules on allowable bend radius and number of bends in a run may not have made sense when you were bending conduit—they will now. These rules are based on the experience of thousands of electricians working over the years, folks who've tried and failed to pull too many wires or go around too sharp a corner. Their trial-and-error efforts have had a direct impact on the code in the form of guidelines to help anyone working with conduit. There are two specialty items that can be of great help when pulling wire through conduit: fish tape and wire-pulling lubricant (*see Steps 2 and 4 below*).

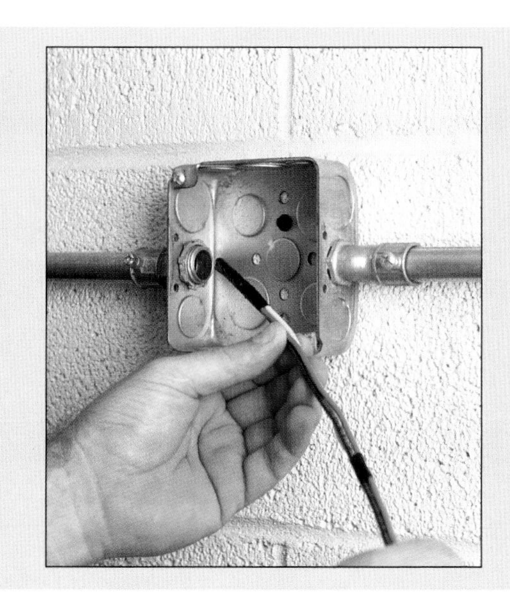

1 **Push for short runs** For short runs where the wire doesn't have many turns to navigate through, it's often possible to push the wire to its destination. One trick I've found helpful is to tape the wire together at the end and about every 6" for the first foot or so. This creates a more solid package that's easier to push. Taping the end also prevents the individual wires from catching on edges of the conduit, connectors, or boxes.

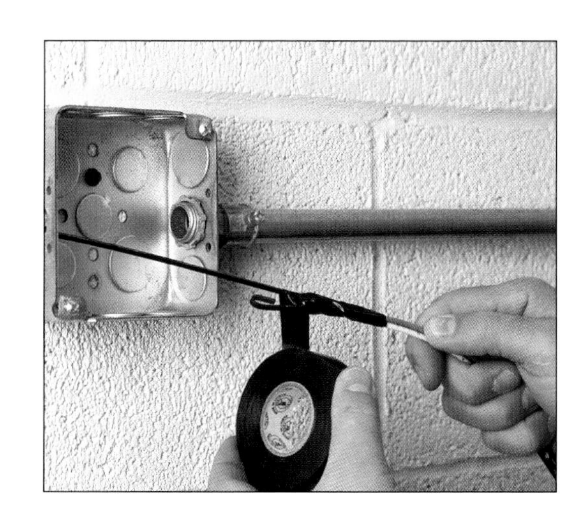

2 **Attach fish tape** For longer runs or when the wire refuses to cooperate, use fish tape. Start by fishing the tape through the conduit to the box where the wire is. Then loop the wires around the hook in the fish tape and wrap this tightly with electrical tape. The smaller the lump, the easier it is to pull the wire. One trick that works well here is to strip about 1½" of insulation off each wire and then loop just the bare copper around the hook.

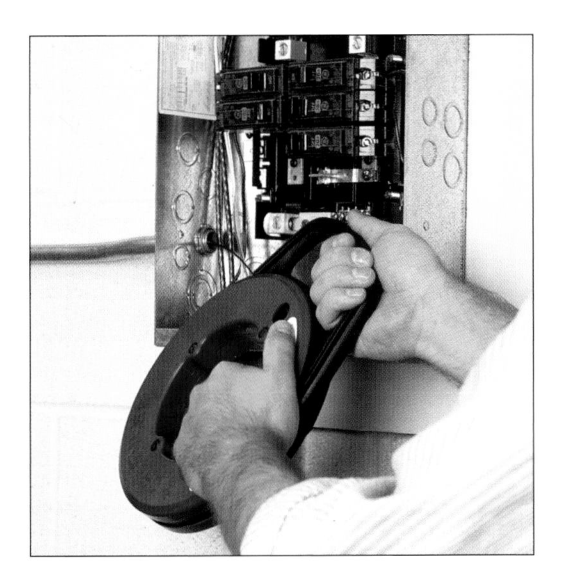

3 **Pull and push the wire for long runs** Once the fish tape is hooked up to the wire, you can get physical. A helper really makes a difference here. The optimal situation is for one person to push and guide the wire into the conduit as the other one pulls. Work slowly with firm, gentle pressure. If you encounter a snag, back up and try again. Jerking on the line will only cause the wire to break, the insulation to shear off, or even the tape to snap.

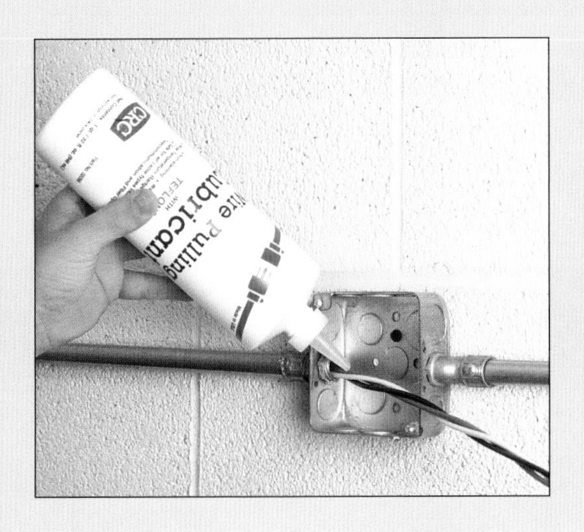

4 **Lubricate the wire or cable** When all else fails, try wire- or cable-pulling lubricant. First, back the wire completely out of the conduit. Then apply a generous amount of lubricant to the wire as it enters the electrical box. Continue applying lubricant until you can feel the wire work past the problem area. At that point you can most likely stop lubricating the cable, since you've got enough in the conduit to do the job.

5 **Leave plenty of excess** After you've successfully pulled the wire or cable to its destination, make sure to leave plenty of excess on both ends. I like to leave 10" to 12" formed into a loop and taped before I cut the opposite end. A classic example of why this is useful is, if you find you can't terminate the line at the breaker that you wanted in a main panel. These extra few inches can save you the trouble of running all new wire.

Surface-Mount Wiring

If you've ever wanted to add a receptacle or a ceiling light and switch but hesitated at the thought of cutting holes in your walls, consider surface-mount wiring. Surface-mount wiring attaches directly to an existing wall or ceiling; it's available in either metal or plastic. Although the metal type will stand up better over time, it isn't as easy to work with as the plastic. Both types are paintable and, when painted the same color as the wall, will almost disappear.

To use surface-mount wiring, you'll first need to "tap" into an existing line; *see Step 1 below.* Then it's simply a matter of running lengths of metal or plastic channel or "raceway" to the desired location. Raceway is available in a variety of precut lengths, or you can cut it to a custom length with a hacksaw (*see Step 2*). Connectors, elbows, and boxes complete the run. Then all that's left is to run wire and hook up the new fixtures.

1 Convert existing box Locate the nearest receptacle where you want your new wiring. Turn off the power to the receptacle, and tape and tag the breaker or fuse panel (*see page 16*). Remove the old cover plate and receptacle. Then attach a "starter" box to the electrical box *as shown.* A starter box has a large rectangular hole in the plate that attaches to the box; this allows the wiring to pass through into the new box.

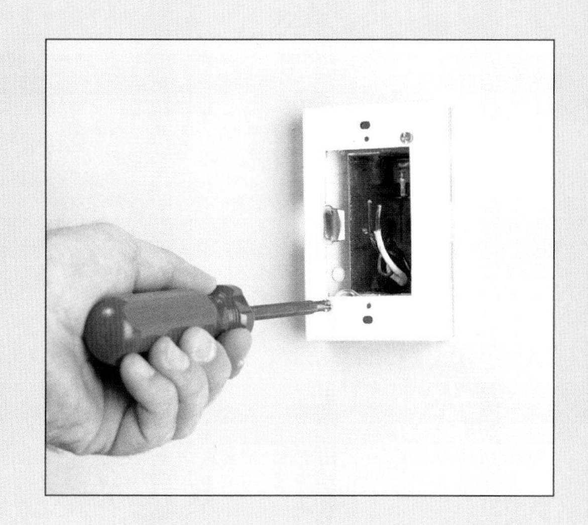

2 Install raceway Now you can measure and cut lengths of raceway to reach the destination of the new box. Press-fit connectors and inside and outside elbows make this an easy task. Just make sure to subtract the length of the connector from the raceway before you cut it to length. Attach the raceway to the wall by screwing directly into studs, or drill holes for plastic anchors.

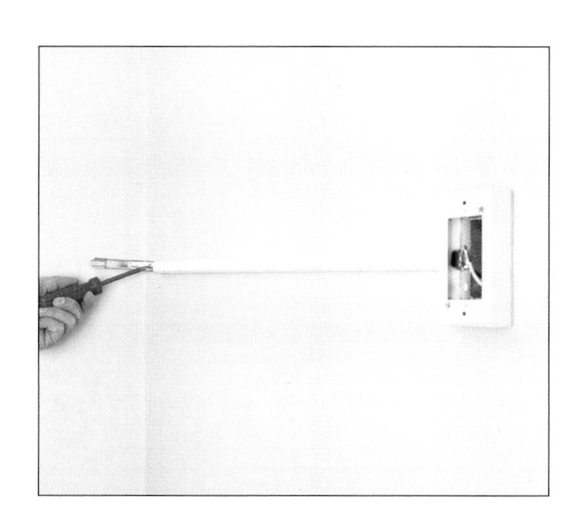

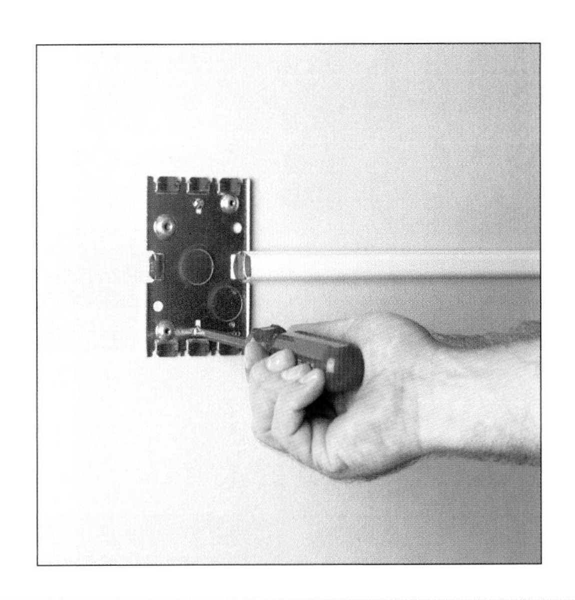

3 **Add boxes** Surface-mount boxes are also attached to the wall either by screwing the plate into studs with wood screws or by drilling holes for plastic anchors. To use plastic anchors, position the back plate of the box where you want it and make a mark through the mounting holes in the back. Then drill holes, insert the plastic anchors, and screw the plate in place.

4 **Run wiring** Running wire through surface-mount raceway is very similar to pulling wire through conduit (*see pages 64–65* for tips on this). Here again, it's useful to tape the ends of the wire together to create a stiffer package—this makes it easier to push the wire through the raceway. Since the 90-degree elbows are so abrupt, it works best to have a helper guide the wire around a corner as you push from the other end.

5 **Attach anchors** Finally, the raceway can be secured to the walls with cable straps (designed especially to fit the raceway). Since the raceway press-fits into the boxes, it's necessary to support it only on long runs and around the connectors (to prevent them from sagging). As before, you can screw these directly into wall studs or (*as shown here*) use plastic wall anchors.

Chapter 4

Emergency Repairs

Picture yourself relaxing comfortably in front of the television, enjoying your favorite evening program. Suddenly the power goes out. What do you do? Panic, if you're not prepared. Relax if you are: You know where the emergency flashlights are; you're familiar with your electrical system and are confident that you can diagnose the problem quickly.

That's what this chapter will go over in detail—preparing for and dealing with emergencies. I'll start by covering the most common problems: tripped or faulty breakers (*pages 69–71*) and blown fuses (*pages 72–74*). Then on to identifying and eliminating shock hazards such as frayed cords (*page 75*) and faulty plugs—everything from snap-on quick-connect plugs for flat or "zip" cords to heavy-duty plugs for large-current-carrying round cords (*pages 76–82*).

Another part of being prepared and identifying problems is knowing how to test switches (*page 83*), receptacles (*page 84*), and GFCI or ground-fault circuit interrupter receptacles (*page 85*). You'll need a couple of inexpensive tools for all of this: a plastic fuse puller if your panel is fuse-based, some sort of circuit tester (such as a receptacle analyzer or probe) to identify receptacle and switch problems, and spare parts (replacement fuses or breakers).

I suggest gathering these supplies into an electrical emergency kit and placing it somewhere that's easy to find. Knowing where your flashlights are is also a big help. If your main breaker panel is in the basement, keep a flashlight inside the door leading downstairs.

Murphy's Law says that the flashlight batteries or bulb will fail when you most need them, so keep spare batteries and bulbs on hand along with a box of emergency candles and some strike-anywhere matches.

When a Breaker Trips

Circuit breakers are really over-current protectors. Normal current flows through the breaker at all times. When current greater than the rating of the circuit breaker flows through it, a bimetal strip inside the breaker heats up and bends. When it bends to a certain point, a spring-loaded contact opens or "trips," effectively cutting off the flow of current. Many good-quality breakers also have a built-in electromagnet to "jump start" the bimetal strip to cut off current flow when the breaker detects a severe short circuit.

Since circuit breakers are heat-sensitive, they can also trip if the ambient temperature inside the panel gets too high. If you've got a breaker (or breakers) that appears to trip randomly, this may be the problem. If leaving the panel door open seems to help the problem, the panel is running hot and it's time to call in an electrician to figure out why.

Tripped Breaker A tripped breaker can be identified in a couple of ways. You can tell the bottom left breaker *shown here* has tripped since the lever is halfway between off and on. Other breakers have a built-in indicator (usually red or orange) that clearly shows when they're tripped. Occasionally a tripped breaker won't be obvious; a quick way to find it is to give each lever a gentle wiggle from side to side. If there's resistance, it's OK; if it wiggles, you've found the tripped breaker.

Resetting a Breaker Most breakers can be reset by first flipping the lever to the OFF position, *as shown,* and then toggling it to the ON position (*inset*). If the problem that caused the breaker to trip still persists, it will immediately trip again. If this happens, unplug any large current users (such as a refrigerator) on the circuit and try again. If it still won't clear, inspect the wiring of the circuit's receptacles, switches, and fixtures.

Replacing a Breaker

Occasionally, circuit breakers go bad and need to be replaced. Or you may want to replace an old breaker that keeps tripping (assuming you've already verified there's not a problem in the circuit). The components inside the breaker—the bimetal strip, the contacts, even the spring—can and do wear out; the breaker should be replaced with an exact replacement if possible.

Manufacturers of circuit breakers make them specifically for their own panels. Because of this, you should use only a replacement breaker designed for your panel. Some manufacturers list on the inside cover of the panel other brands that have been tested and found suitable as replacements. When you go to buy a new breaker, jot down the replacement names as well as the maker of your panel—you'll have better luck finding what you need.

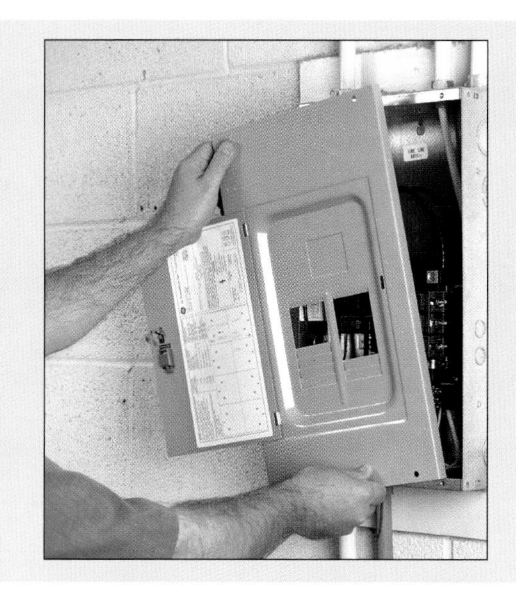

1 Remove panel cover Replacing a breaker isn't difficult, but you need to pay close attention to the safety rules for working inside an energized panel (*see pages 16–19*). The first thing to do is remove the panel cover; it's typically held in place with four slotted screws. After you've removed the screws, lift the cover out and off and set it aside. (I usually thread the screws right back into the panel so I won't misplace them.)

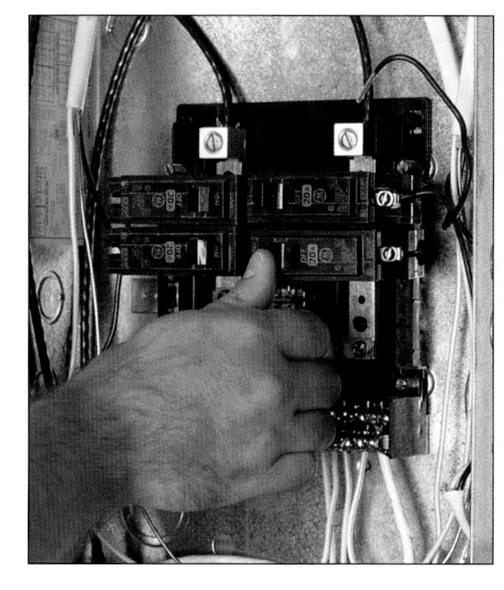

2 Turn off the breaker Although it would be best to turn power off to the panel completely, this usually isn't possible. If you can't, turn off the main breaker for the panel. At the very minimum, turn the power off to the breaker that you need to replace.

3 **Disconnect wiring** If you're working inside an energized panel, be careful and take your time. There's only one wire to disconnect from the breaker: the black wire connected to the load terminal. Using an insulated-grip screwdriver, reach in with one hand and loosen the screw. Set the screwdriver down and, with the same hand, reach in and carefully pull out the black wire. Don't touch the bare end; temporarily twist a wire nut on the end to prevent accidental contact.

4 **Replace breaker** Most breakers are held in place by the contacts that slip over the panel bus; they're located under the breaker closest to the center of the panel. A clip under the load terminal screw on the other end hooks onto the panel. To remove a breaker, grasp it firmly *as shown* and pull up on the end near the center. Press the new one into place by first hooking the clip onto the panel; then press the tabs onto the bus.

5 **Reconnect** With the breaker off, remove the wire nut that you temporarily twisted on the end of the black wire. Carefully slip this wire into the load terminal. Then use an insulated-grip screwdriver to tighten the screw; make sure it's good and tight. Gently pull on the wire to make sure you have a solid connection. Replace the main panel cover and restore power.

Replacing Blown Fuses

Most service panels installed prior to 1956 have fuses instead of circuit breakers. Screw-in type fuses protect 120-volt circuits; cartridge-type fuses (*page 73*) protect 240-volt circuits. Current flows through a thin metal strip inside the fuse. When too much current flows, the strip melts to stop the flow of current. A close examination of the fuse will tell you what kind of a problem occurred: A cleanly melted strip indicates a current overload (too many devices demanding current at the same time; e.g., the dishwasher and microwave both at once). If the fuse window is discolored, it's a sure sign of a short circuit and that there was an almost instantaneous demand of current—enough to blow up the strip instead of melting it. If the problem isn't fixed, the fuse will immediately blow again. If this occurs, it's time to call in an electrician.

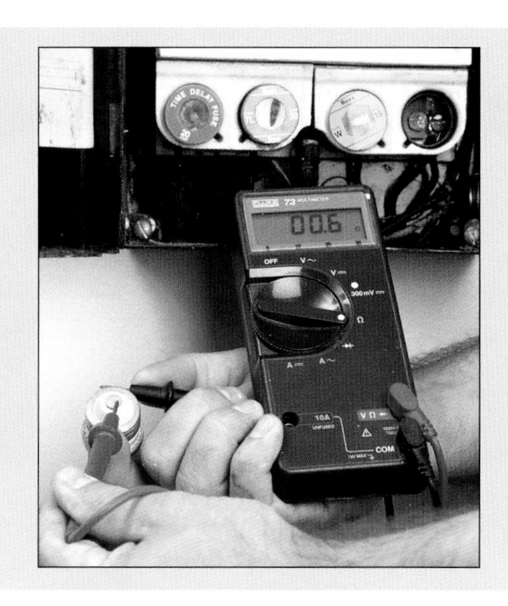

1 **Identify blown fuse** Sometimes a blown fuse isn't apparent. Occasionally the metal strip melts where you can't see it. If you aren't sure a fuse has blown, check it with a continuity tester or a multimeter. First, turn off the power to the panel and unscrew the suspect fuse. Switch the meter to measure resistance (ohms), and touch one lead to the rim of the fuse and the other to the contact on the bottom. A good fuse will measure almost zero.

2 **Unscrew old fuse and replace** If it's clear which fuse has blown, turn of the power to the panel, if possible, and carefully unscrew the fuse. Make sure to touch only the insulated portion of the fuse and not the metal rim. Fuses should always be replaced with one of identical amperage rating. Never replace a blown fuse with one of a higher rating in an attempt to prevent the fuses from blowing!

Replacing Cartridge Fuses

Cartridge fuses are used to protect 240-volt circuits. Unlike their screw-in–type cousins, cartridge fuses don't provide a visual indication that they have blown. This means you'll need to remove the fuse to test it (for more on this and a trick for testing a fuse in circuit, *see the sidebar on page 74*). I've also encountered cartridge fuses that "partially" blew; that is, they allowed current to flow, but not the correct amount. Cartridge fuses make contact with the service panel via a set of spring clips. Fuses for 30- to 60-amp circuits usually have ferrule-type ends (*as shown on page 45*). Fuses with knife-blade contacts are designed to carry 70 amps or more. Some cartridge fuses fit in compartments that snap into the service panel (*see page 12*), while others insert directly into the panel (*as shown below*).

1 Shut off power Many smaller auxiliary cartridge fuse service panels have a separate power shutoff in the form of a large lever on the outside edge of the panel. Some panels are designed so that you can't open the panel door until this lever is switched to the OFF position. If you're working on a larger main service panel, turn the main power switch to the OFF position.

2 Remove fuse with fuse puller Use a plastic fuse puller to remove the suspect fuse. Grasp the middle of the fuse with the puller, and pull it out of the spring clips. If it has been in there awhile, this can take a surprisingly hard jerk to free it. Be careful not to touch the metal spring clips—they might still carry current. Set the fuse on an insulated surface—it may be hot. Let it cool before testing it with a multimeter.

3 **Replace with new fuse** Even if you've shut power off to the panel, I don't recommend installing a new cartridge fuse by hand. Instead, insert the new fuse by holding it firmly with the fuse puller and pushing it sharply into the spring clips. Seat the fuse properly in the clips to ensure that it is making good contact, then give it a gentle tug to check that the spring clips are holding it securely.

TESTING CARTRIDGE FUSES

There are two ways to test a cartridge fuse to see if it's good: out-of-circuit and in-circuit. The safest way is to test the fuse after it has been removed from the panel with a fuse puller. With the fuse out of the circuit, place the leads from a multimeter or continuity checker on the metal ends of the fuse as shown *below left.* A good fuse will measure near zero (the lamp on a continuity checker will light). If the lamp doesn't light or the meter reads infinity, the fuse is bad.

An in-circuit test can be made by switching the multimeter to read AC voltage and touching one lead of the multimeter to each end of the fuse (*below right*). A good fuse will read 0 volts; a blown fuse will read 120 or 240 volts. This method is particularly useful for finding partially blow fuses or bad or corroded spring clips. Caution: Make sure the meter is set to read voltage—if it's set to read amperage, you'll damage the meter.

Repairing Frayed Cords

It's a good idea to inspect the electrical cords in your home regularly. Whenever you come across a cord where the insulation has been abraded or torn, it needs to be replaced, or else cut at the worn spot and a new plug put on. The repair should be made as soon as possible. If you can't get to it immediately, unplug the device and set it aside. If you must continue to use the device, protect your family from electric shock with one of the two temporary fixes below.

When it comes time to repair the cord, inspect the entire cord. In instances where the cord is long or the damage is near the plug, you can cut the cord on the device side of the damage and simply install a new plug. *See pages 76–78* for a standard plug, *pages 79 and 80* for quick-connect plugs, and *page 81* for an industrial-style plug.

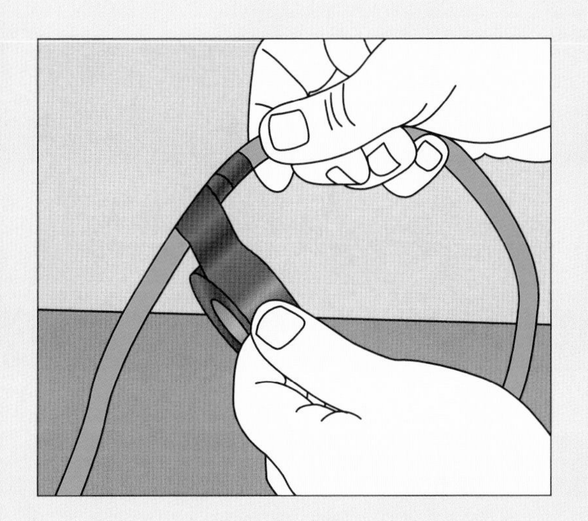

1 Temporary fix: electrician's tape A quick (but only temporary) way to fix a frayed cord and a cord with missing insulation is to wrap a few turns of electrician's tape around the damaged area. Start a few inches in front of the damaged area and continue wrapping for a few inches past the damage. Remember that this is only temporary—the tape will eventually unwind, leaving the cord unprotected.

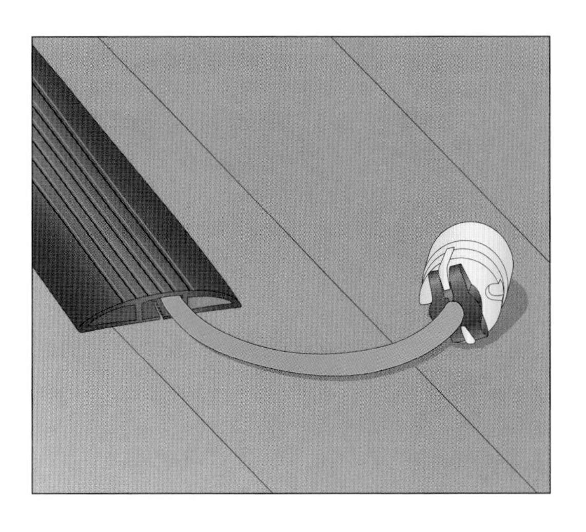

2 Temporary fix: cord cover If the cord is in a heavy-traffic area such as a family room, a temporary solution is to insert the cord in a cord protector. These extruded rubber pads are designed to hold and protect cords. They're often used in industry where people and machinery travel on top of cords. You can purchase these from office supply stores, from some home centers, and from safety equipment suppliers (look in the yellow pages under "safety equipment").

Replacing a Standard Plug

One of the most common shock hazards in the home is a faulty plug. The primary reason they fail so regularly is that they take so much abuse. Plugs get stepped on, rolled over, bumped against, and yanked by the cords. To repair a plug, you'll first need to identify the type of cord and the style of replacement plug you'll need to purchase. Flat cords (such as the one *shown here*) are commonly referred to as zip cords and are made up of two conductors that are encased in insulation that is easily "zipped" apart. This cord, suitable for lamps and other low-current devices, will accept a standard plug or quick-connect plugs (*pages 79–80*). Round cords, with their thicker conductors, can handle more current and are typically terminated with a polarized and grounded plug.

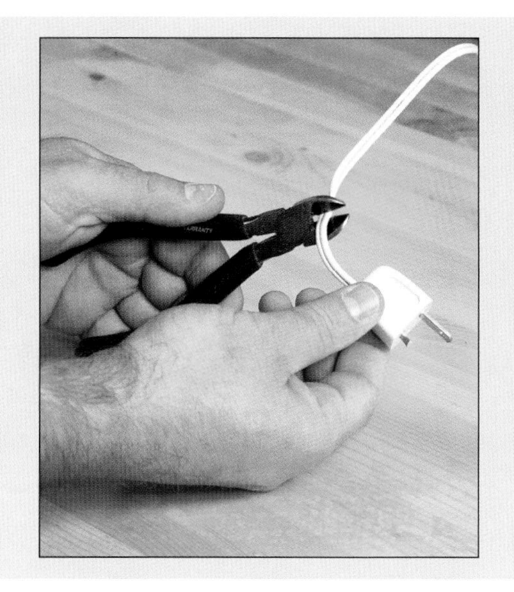

1 **Remove old plug** To replace a plug, start by removing the old one. On encased plugs (*as shown here*) you'll need to cut the plug off. To maintain as much length as possible, cut the cord as near to the plug as possible. If the cord is damaged, cut it on the device side of the damage. For screw-type plugs, remove the insulator, loosen the screw terminals, disconnect the wires, and lift the plug off the cord.

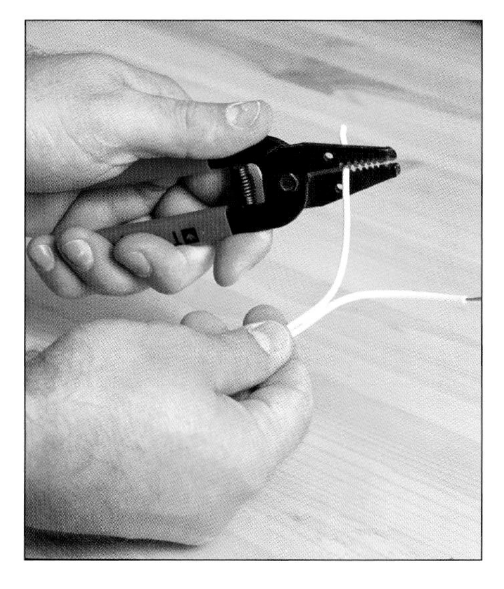

2 **Strip insulation** When working with flat or "zip" cord, separate the conductors about 2" and remove about ¾" of insulation from each end with wire strippers. If you're reusing a round cord, check the condition of the wires. If they're tarnished or broken, cut the ends off and restrip (you may need to remove some of the outer insulation to expose sufficient lengths of the inner conductors).

3 Insert wire in new plug Remove the insulating cap from the new plug (if it has one) or pull the termination end of the plug out of the plug housing. Then carefully pass the stripped end of the cord through the plug housing. (This may seem obvious, but I can't tell you how many times I've been in a hurry to hook up a new plug, only to find out that I forgot to first slip on the protective housing.)

4 Underwriters knot For plugs without built-in strain relief (like the round plug *shown on this page*), and for added pullout protection for even for those plugs that do, tie the conductors together with a special Underwriters knot *as shown.* After you've tied the knot, pull the ends of the conductors to tighten the knot.

5 Pull knot down into housing Check to make sure the knot will do its job as a strain reliever by pulling the knot down and into the plug housing. If you can still pull it through, you'll need to purchase a small clamp to fit around the section where the cord passes out of the housing. Tightening the clamp will compress the sleeve and grip the cord.

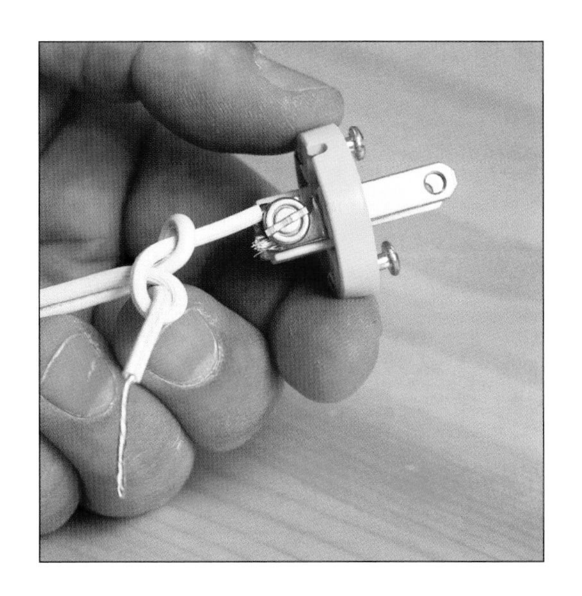

6 **Wrap around screws** Twist the ends of wire tightly together and then wrap each conductor around the appropriate screw terminal. (If you're working with round cord: The black wire wraps around the brass screw, the white wire goes around the silver screw, and the green goes around the grounding screw.) Wrap the conductors clockwise around the screws so they'll pull tighter as the screw is tightened.

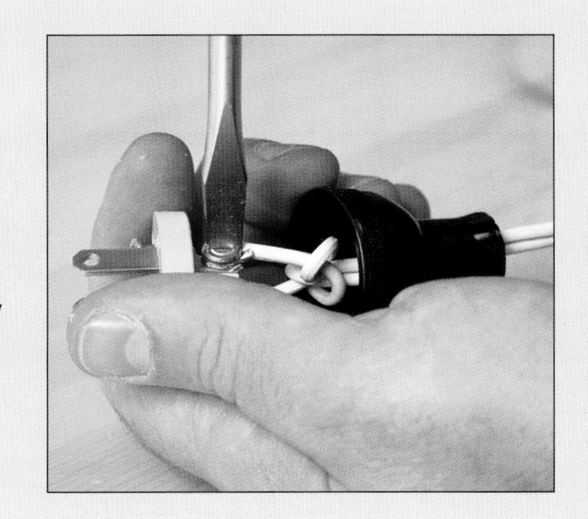

7 **Tighten screw terminals** Use a screwdriver to tighten the screws. Don't be tempted to overtighten these—all you'll accomplish is squishing the twisted strands. When this occurs, a few strands of the wire can lift out and create a potentially dangerous situation: They can reach across to one of the other conductors (especially if you overtighten all the screws) and cause a short circuit.

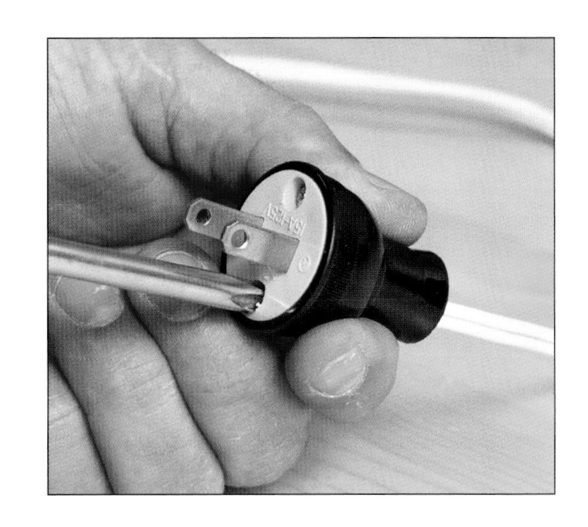

8 **Assemble the plug** All that's left to do is assemble the plug. On some plugs, all this entails is snapping or screwing on an insulator over the prongs to cover the exposed wiring. On plugs like the one *shown here,* you insert the plug into the housing and then tighten the mounting screws. If the plug has a strain-relief clamp, tighten it as well.

Installing a Quick-Connect Plug

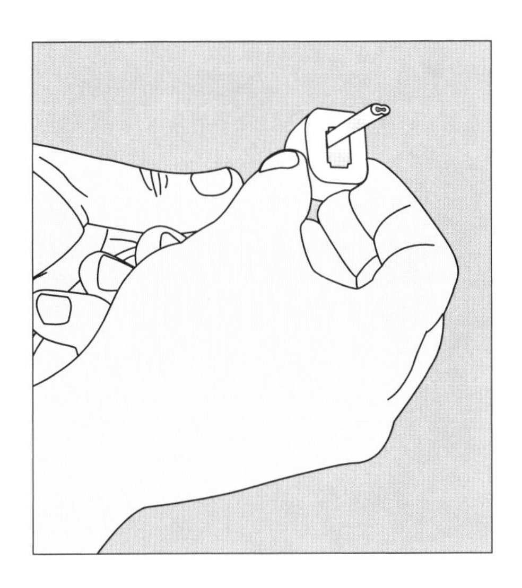

1 Prepare the cord Install a quick-connect plug on zip cord by first cleanly cutting the cord with wire cutters. Then squeeze together the plug prongs, and pull to remove the inner unit from the cover. Set the inner unit aside and pass the cut end of the zip cord through the plug cover.

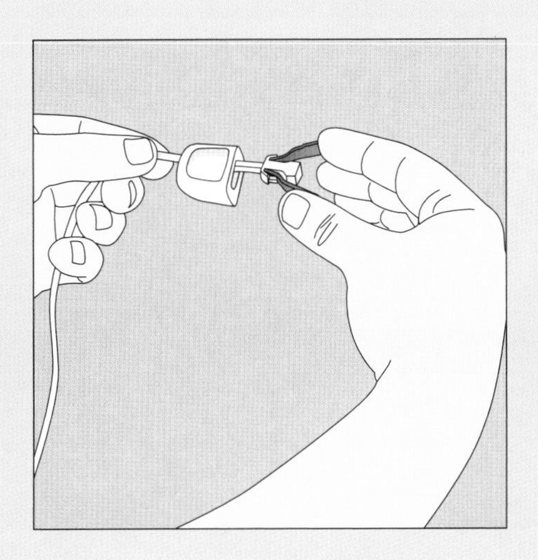

2 Insert in plug To ensure proper polarization, hold the inner unit so the wide blade is up and the cord has the ribbed side to the right (see the manufacturer's directions on the package). Note: If neither side of the cord has ribs, the side with the white/silver conductor should be considered the ribbed side. Press the unstripped end of the cord firmly all the way into the inner unit.

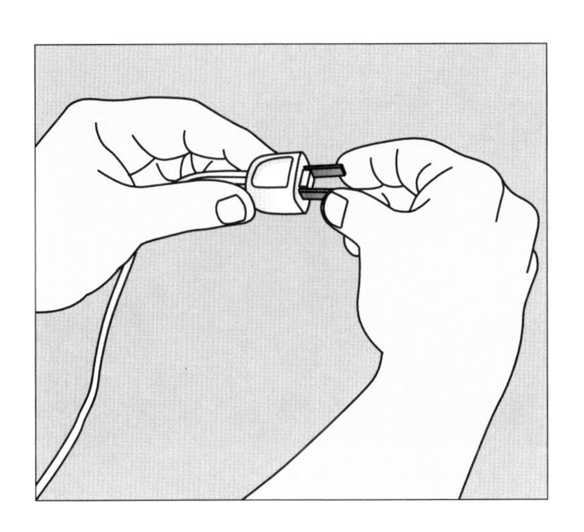

3 Snap together Now you can squeeze the prongs together. This forces small V-shaped teeth to pierce the insulation and make a solid connection to each conductor. Press the squeezed-together inner unit into the plug cover. You'll know when the inner unit is in far enough because you'll feel a slight click as mating plastic edges engage. Give the cover a gentle tug to make sure that it holds fast.

Installing a Quick-Wire Plug

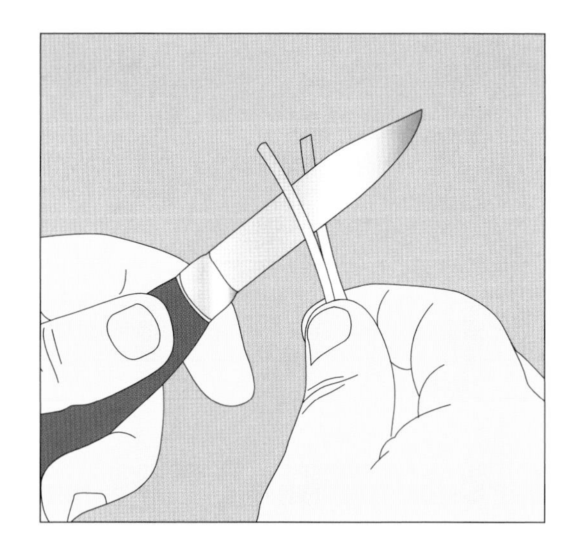

1 Prepare the cord Prepare the end of the cord for the quick-wire plug. For zip cords, cut the end cleanly and then separate the two conductors about 1¼" *as shown.* For round cords, cut and peel back 1¼" of the outer cord insulator. Do not strip off the insulation from the individual conductors.

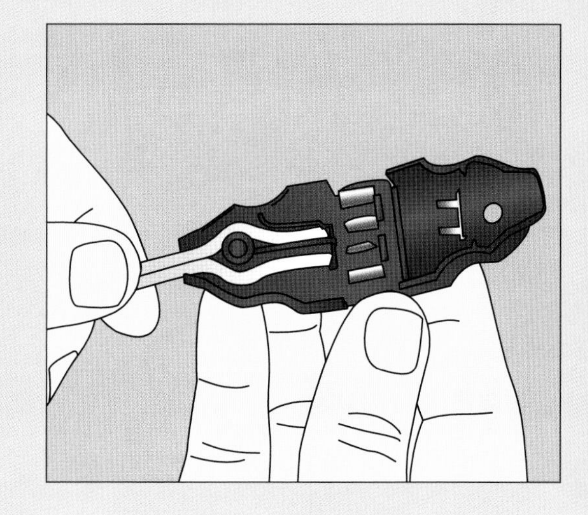

2 Insert conductors Open the plug carefully to the flat position. Place the cut ends of the cord in the channels in the plug end *as shown.* Then push the ends of the cords up against the stops at the ends of the channels. Make sure the conductors pass cleanly around the post and butt firmly against the stops.

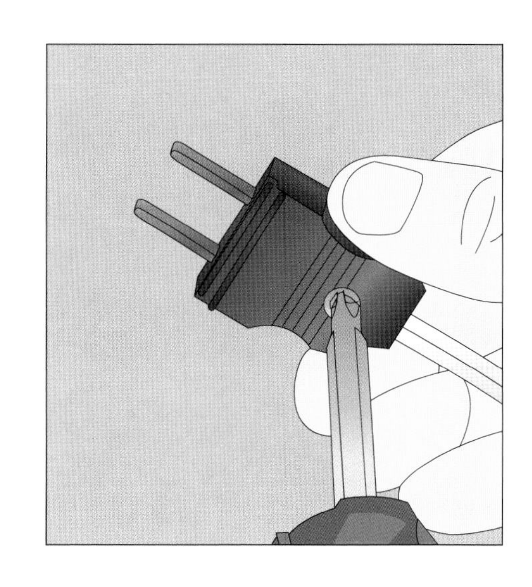

3 Assemble the plug Now pivot the cover up and onto the plug side. Press the two pieces together firmly to force the metal contacts on the cover side into the conductors to make a solid connection. Drive in the screw to close the housing completely, and then secure the assembly. Test the strain relief by pulling gently on the cord: There should be no movement at all.

Installing a Heavy-Duty Plug

Heavy-duty plugs are known by many names often describing their end use: commercial, industrial, even hospital. What they all have in common is a rugged casing with a built-in strain relief designed to stand up to prolonged use and abuse. They're ideal for cords that get plugged and unplugged often—like a shop vacuum, electric tools, heavy-duty fans, and so on.

The casings often are made of high-impact plastic and come in two parts that screw together. Inside a heavy-duty plug there are built-in clamps that accept the wires in lieu of stripped wires wrapping around screw terminals. When the corresponding clamp screw is turned, the wire is gripped securely—this makes this type of plug quick and easy to install.

1 Remove insulation and strip wires To install a heavy-duty plug on the end of a round cord, start by cutting about 1¼" of the outer cord insulation off with a pocketknife or utility knife. Be careful not to nick the insulation of the inner conductors. Then strip about ½" of insulation off the end of each conductor with a pair of wire strippers.

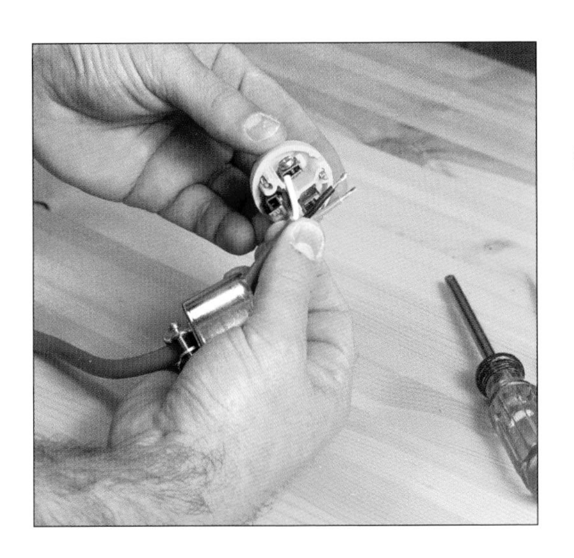

2 Insert wires Before you connect the cord to the plug, slip the plug housing over the cord *as shown.* Next insert the stripped end of each wire in the small clamp for its respective terminal: black wire to brass, white wire to silver, green wire to green. Make sure you insert the wire between the C-shaped cap and the corresponding terminal.

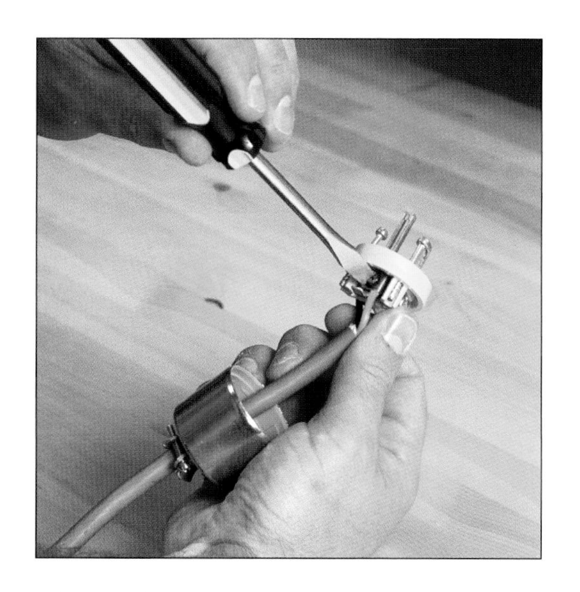

3 **Screw wires to terminals** Now you can tighten the screw for each terminal clamp to secure the wire. Here again, it's important not to crush the wire. Overtightening will only cause strands of the wire to squish out from under the clamp. And since there's not a lot of room inside one of these plugs, it's easy for one of these errant strands to make contact with another wire, causing a short.

4 **Screw together the halves** After you've tightened the screws, slide the plug housing up the cord until it mates with the other half of the plug—press gently and make sure the wires don't get pinched as you position the housing. Often there is an alignment tab to help make sure that the plug goes together correctly. Once together, use a screwdriver to tighten each of the screws that hold the plug together.

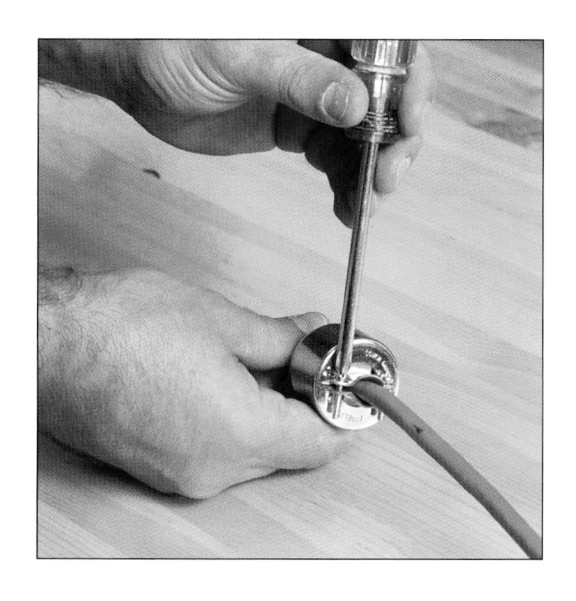

5 **Tighten the strain relief** The last thing to do is tighten the screw(s) for the clamp that provides strain relief. Some cords have a rubber bushing inside the clamp to protect the cord from the clamp. Others simply squeeze the cord to prevent it from moving. In either case, tighten the screws only enough to capture the cord; overtightening will only compress the conductors, which can cause future problems.

Testing Switches

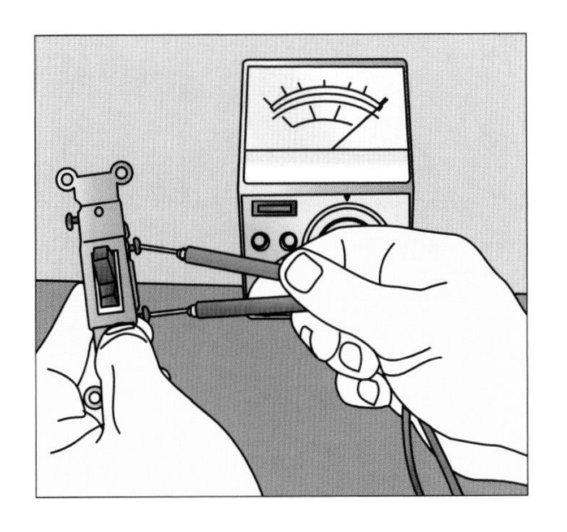

Single-Pole Testing a single-pole switch is simple: Just touch one lead of a multimeter or continuity checker to one screw terminal and the other lead to the other screw terminal. If you're using a multimeter (*as shown here*), switch it to read resistance (ohms). Toggle the switch lever from on to off. The reading on the multimeter will change from zero to infinity if the switch is good; the lamp in a continuity checker will go on and off.

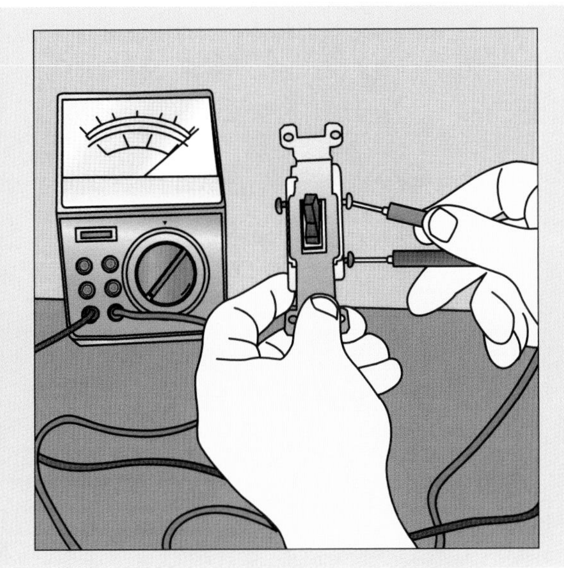

Three-Way Attach one lead of the tester to the black common terminal. Then touch the other lead to one of the traveler terminals. Toggle the switch back and forth; the tester should indicate a change (the lamp goes on and off or the reading changes from zero to infinity). Now move the lead to the other traveler terminal and repeat the test. Again, you should see a change, but this time when the switch lever is in the opposite position.

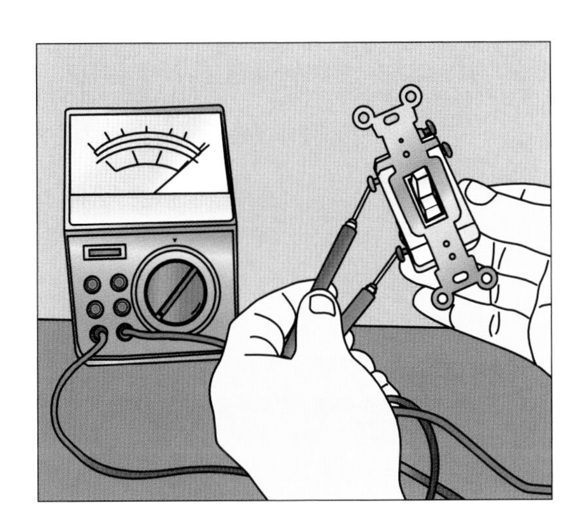

Four-Way Testing a four-way switch can be a bit confusing because the arrangement of the pathways inside the switch will vary depending on the manufacturer. What you're looking for is four continuous pathways between the screw terminals—two for each lever position. Touch the tester leads to each pair of screw terminals in turn, vertically, horizontally, and diagonally; and toggle the switch lever to test the switch.

Testing Receptacles

Neon Circuit Tester To check for power with a neon circuit tester, insert one probe in each slot of the receptacle. If the lamp glows, you've got power. You can also check for proper grounding. For a three-slot receptacle, insert one probe in the short (hot) slot and the other in the ground—the lamp should glow. For a two-slot receptacle, insert one probe in the short slot and touch the other probe to the cover plate screw—it should also glow.

Multimeter Unlike the circuit tester *above,* a multimeter, also called a multitester (or, technically, a VOM: volt-ohm-milliammeter), will not only check for power, it'll tell you exactly how much is there. To use a multimeter to check power, first switch it to read AC voltage and pick a range above the receptacle's voltage (the meter *shown here* auto-ranges—it automatically selects the best range). Then holding the test probes by their insulating handles, insert one probe in each receptacle slot and take a reading.

Receptacle Analyzer A receptacle analyzer is a handy tool that, in addition to verifying power at a receptacle, will also check for numerous wiring faults. Typically, the analyzer will let you know if there's a reverse polarity problem or a bad grounding connection. To use one, simply insert the analyzer into the receptacle, note which of the three lamps is lit, and check the chart on the analyzer to assess the condition of the receptacle.

Testing GFCI Receptacles

A ground-fault circuit interrupter (GFCI) has a built-in microprocessor that constantly checks for small leakages of current that indicate a ground fault. If the circuit is grounded properly, these leakages aren't a big problem. But if you are well grounded (standing in a puddle of water, e.g.), the current could pass through you instead of the electrical system. Here's where the GFCI outlet comes into play. It monitors both legs of a circuit to make sure they're the same. If it detects more current flowing in one leg than in the other, it disconnects the circuit. That is, as long as it's operating properly. The fact that you have a GFCI outlet installed (*see page 96*) doesn't mean it's working. That's why GFCI manufacturers build in a self test. GFCI outlets should be tested at least once a month—and it only takes a few seconds (*see below*).

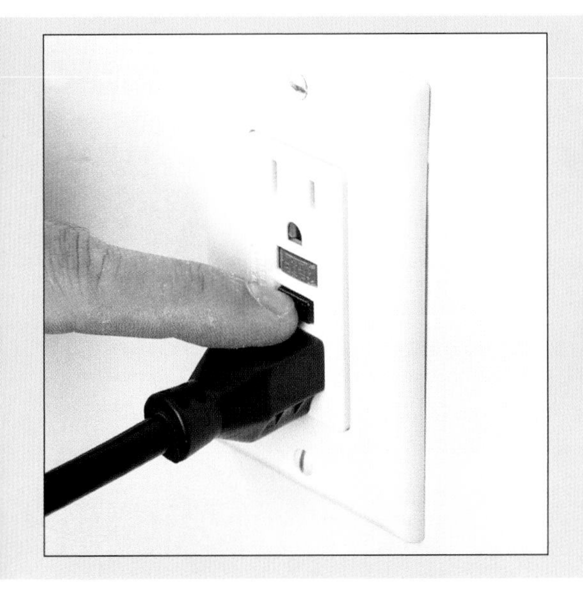

1 Press test To test a GFCI receptacle, press the test button. It's a good idea to have a device plugged into the receptacle to check to see if it's working. If the GFCI receptacle is wired to protect additional receptacles, make sure you plug in a device in each of the protected receptacles as well.

2 Reset should pop out When you press the test button, the reset button should instantly pop out, and power to the device plugged into the receptacle should disconnect. This extra step of plugging in a device can save a life in a ground-fault situation. I've seen GFCI receptacles that tested fine (the reset button popped) but failed to disconnect power. They are, after all, electronic devices that can (and will) eventually wear out (just like your TV or VCR).

Chapter 5

Working with Boxes, Receptacles, and Switches

Since code requires that all wire connections or cable splices be made inside a box, you'll come across electrical boxes on virtually every project or repair you tackle. Besides holding wire for connections and protecting framing members from sparks that could potentially start a fire, boxes also protect two of the most common electrical devices in your home: receptacles and switches. Both of these are subjected to constant use (and sometimes abuse)—and will fail over time and need to be replaced.

In homes built prior to the 1960s, it was common to install ungrounded receptacles. Most electrical codes now require that grounded receptacles be used to replace ungrounded receptacles. The exception to this is a home with a two-conductor wiring system. In this situation, you can replace an ungrounded receptacle with a GFCI receptacle to obtain ground-fault protection (*page 96*).

In this chapter, I'll start with detailed instructions on installing boxes—both in new construction and in preexisting walls—to accept receptacles (*pages 87–88*), light fixtures (*page 89*), and switches (*page 90*). Then I'll show you how receptacles are commonly wired and how to replace them (*pages 93–95*). (See *page 84* for instructions on how to test a receptacle.) Next is upgrading or replacing a ground-fault circuit interrupter (GFCI) receptacle (*pages 96–97*)—a common code requirement for various locations in your home.

After that, I'll unravel the mysteries surrounding switch wiring—single-pole, three-way, and four-way switches—how they're wired, along with detailed instructions on how to replace them (*pages 98–102*). (See *page 83* for directions on how to test switches.) And finally, I'll give step-by-step instructions on how to replace a simple on/off switch with a dimmer switch so that you can vary a light fixture's brightness (*page 103*).

Installing Boxes for Receptacles

Before you begin any work installing electrical boxes, have your project plans approved by an inspector—all boxes must be accessible (you can't bury one inside a wall). Then you can use any of the electrical boxes shown on *page 34* to install receptacles quickly when the wall studs are exposed. Most of these new construction–style boxes have a built-in gauge to make it easy to position the box on the studs so the box will end up flush with the finished wall.

Whenever possible, use the deepest boxes that are practical so you'll be sure to meet code requirements for box volume (*see pages 30–32*), and also to give you more "knuckle" room to make your wiring connections. Deeper boxes also allow for future expansion. It's also a good idea when you install electrical boxes in adjacent rooms to position boxes that share a common wall close together. This simplifies wiring and uses less cable.

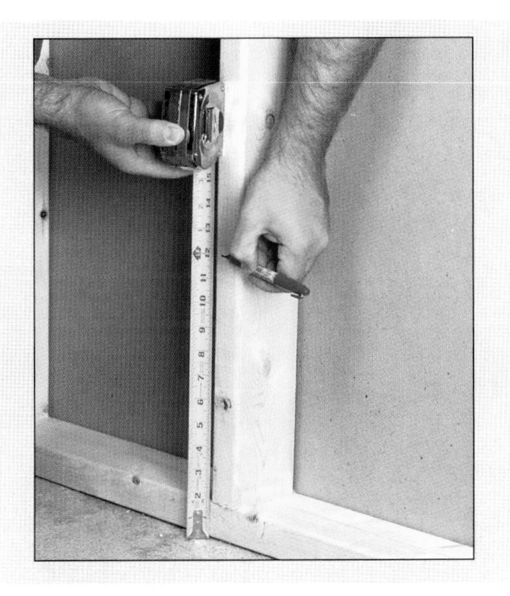

1 Mark the location To install an electrical box for a receptacle, the first thing to do is measure up from the floor the correct height and make a mark. Most codes specify that this distance be 12" to 16" above the floor. Check your local code to make sure you meet its requirements. GFCI receptacles in bathrooms and kitchens should be mounted so they're 10" above the countertop.

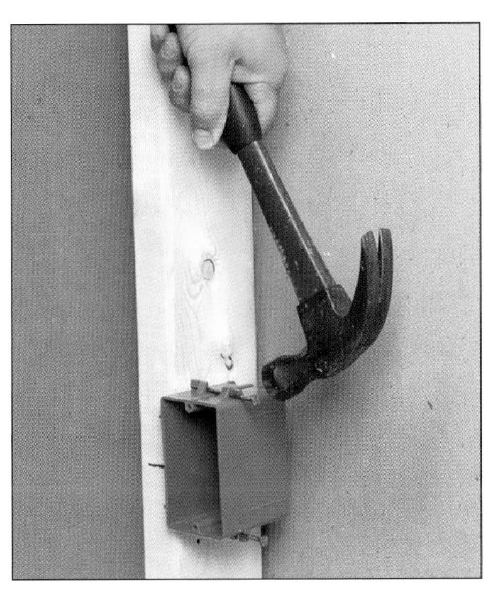

2 Position the box and nail it Center the box on your mark and set it out the correct distance so it'll end up flush with the finished wall (*see page 90*). If your box doesn't have a built-in gauge or marks, hold a piece of the finished wall material up against the stud and slide the box forward until it's flush. Then hammer in the fasteners to secure it to the stud, and prepare the knockouts (*see page 88*).

Plastic Tabs Most plastic new-construction boxes have built-in tabs that are snapped out to accept a cable. The quickest way to remove a tab is to pry it out with a flathead screwdriver. Just insert the blade of the screwdriver in the tab you want to remove, and give it a quick snap. Use your screwdriver to break off any sharp edges that remain so that non-metallic cable won't get damaged as it's pulled though the slot.

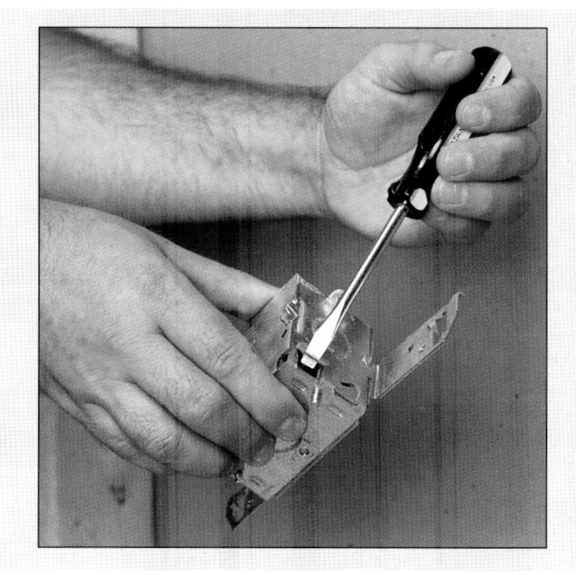

Pry-Outs Some metal electrical boxes have slotted pry-outs that can be quickly removed to make way for a cable. Poke the tip of a flathead screwdriver in the slot of the pry-out you want to remove, and give it a good twist as you pry up. Most boxes with pry-outs have interior box cable clamps to prevent the cable from moving and abrading the cable sheathing.

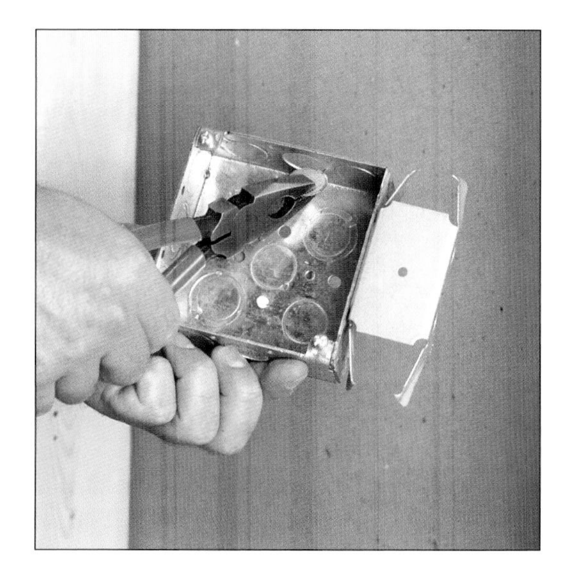

Knockouts The most common form of cable access into a metal box is the knockout. These are holes that are punched almost completely though the box during manufacture. A small section is left attached to the box to hold it in place. Remove a knockout by first tapping it with a hammer or linesman pliers. Then grasp the knockout from inside the box with the pliers and twist it to break it off.

Installing Boxes for Light Fixtures

Position boxes for ceiling fixtures in the center of a room. When there's access to joists, this is simply a matter of finding the right joist and installing the box; *see below.* If you have access above the ceiling (such as unfinished attic space), just locate the center and cut a hole in the ceiling to match the box you're using; then install a cleat to support the box.

Vanity lights can be mounted in boxes above a mirror, but the best lighting scheme is achieved when two lights are installed on each side of a mirror. Some light fixtures don't require a separate box: recessed fixtures that fit inside wall cavities, bathroom fans with lights, and surface-mounted fixtures such as under-cabinet fluorescent lights. These fixtures have built-in boxes and usually aren't installed until final connections are made.

Position on Stud New-construction boxes for light fixtures are designed to be hammered into the joists. To install one, position the box where you want it and such that it will be flush with the finished ceiling. Hammer in the mounting nails to secure it. Many light fixtures will require that you attach a mounting strap to the box—these typically come with the fixture.

Bracket for between Studs To position a light fixture between ceiling joists, attach the box to the brace bar and adjust the bar to span the joists. Nail the ends of the brace to the joists *as shown* so that the face of the box will be flush with the finished ceiling. Then slide the box along the brace to the desired position and tighten the mounting screws.

Installing Boxes for Switches

1 Locate the box Before installing an electrical box for a switch, check your local code for the correct height. Switches are usually mounted on the latch side of a door, with the box 48" from the floor. For areas where portions of the walls are tiled, boxes are typically mounted 54" to 60" above the floor so that they won't interfere with the tile.

2 Nail the box in place Once you've located the box, hold it in position so the front face will be flush with the surface of the finished wall (*see below*), and nail or screw the box to the wall stud. Then remove the appropriate knockouts for cable (*see page 88*). Note: You can mount a box between studs by first installing a cross block (a scrap of 2×4) between the studs. Then attach the box to the cross block.

Finished Wall Depth

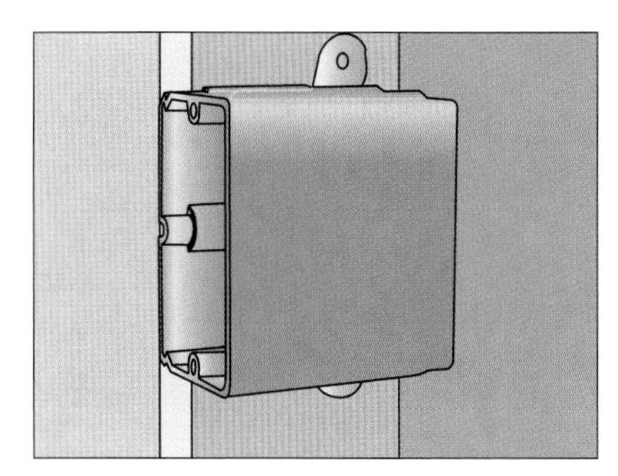

Drywall Attach box so front face extends the thickness of the drywall past the wall stud.

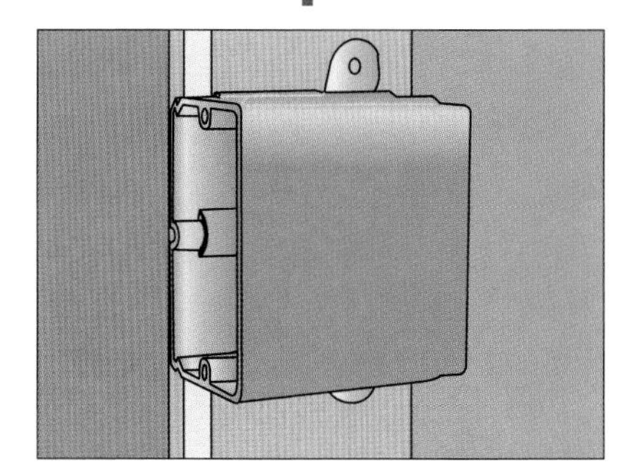

Paneling over Drywall Attach box so front face extends 1/8" plus thickness of drywall past the wall stud.

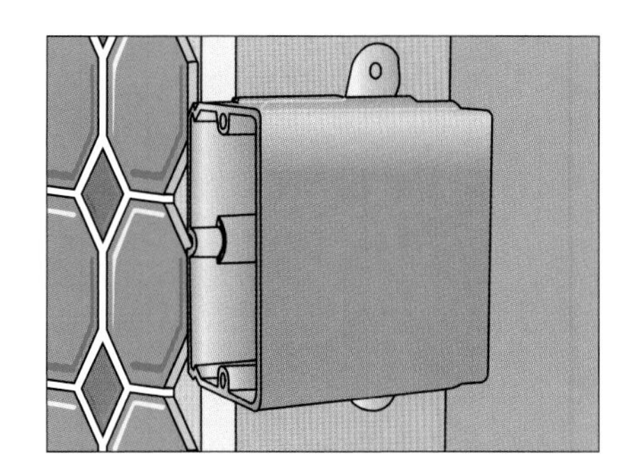

Tile over Drywall Extend boxes 7/8" past the wall studs (adjust for thicker or thinner tile).

Installing Boxes in Preexisting Walls

If you're planning on extending a circuit (*page 117*) or adding a new light fixture (*page 114*), you'll need to install an electrical box in a preexisting wall or ceiling. Although the thought of cutting a hole in your wall or ceiling might make you nervous, it's really a simple process. (If you don't feel up to this, consider surface-mount wiring; *see page 66.*) The only tricky part is positioning the hole so that it doesn't interfere with the framing. Fortunately, a simple tool called a stud finder will prevent this from happening (*see Step 1 below*).

Electrical box manufacturers have made installing boxes in preexisting walls easy by designing numerous types of boxes that don't have to attach to the framing (*see page 35*).

1 Locate studs The most reliable way to locate wall studs or ceiling joists is with an electronic stud finder like the one *shown*—they cost around $10 each and can be found at any hardware store or home center. Alternatively, you can drill a small hole in the wall and insert an L-shaped piece of wire; rotate the wire until it hits a stud, and make a mark.

2 Mark box outline Measure up from the floor the correct height for your box (*see page 87* for receptacles, *page 89* for light fixtures, and *page 90* for switches) and make a mark. Then use the box you're installing as a template. Hold it in position (away from the stud and at the correct height) and trace around it. Some box manufacturers include a handy template with their boxes to make this even easier.

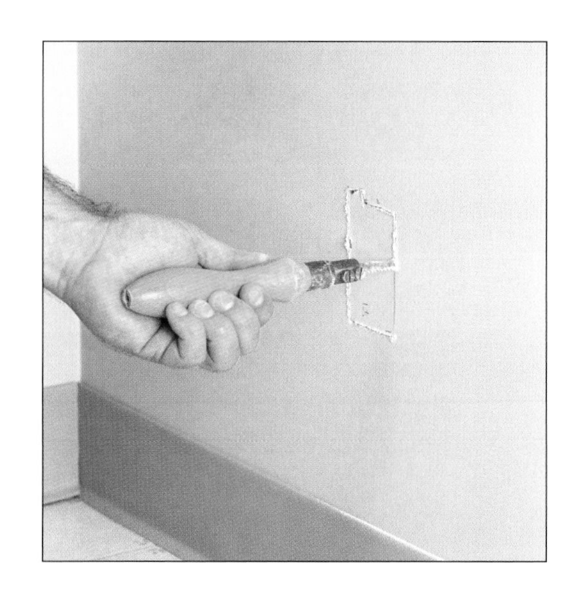

3 **Cut holes in the wall** Now comes the fun part: cutting a hole in the wall. For drywall, use a drywall saw; push the tip of the blade through the drywall at a corner and cut to the opposite corner; repeat for all four corners. For lath and plaster, chisel out the plaster inside the lines of the marked hole, drill 1"-diameter holes in the corners, and cut the lath with a keyhole saw or a reciprocating saw.

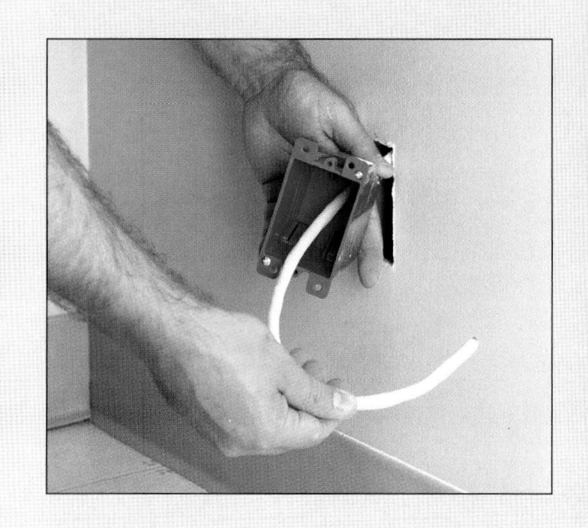

4 **Run cable into the box** Feed the new cable up into the hole you just cut. (*See pages 120 and 121* for more on running cable through walls and ceilings.) Remove the appropriate knockouts (*see page 88*), and insert and secure the cable to the box. Make sure to leave yourself plenty of excess cable for making connections to the switch, receptacle, or fixture.

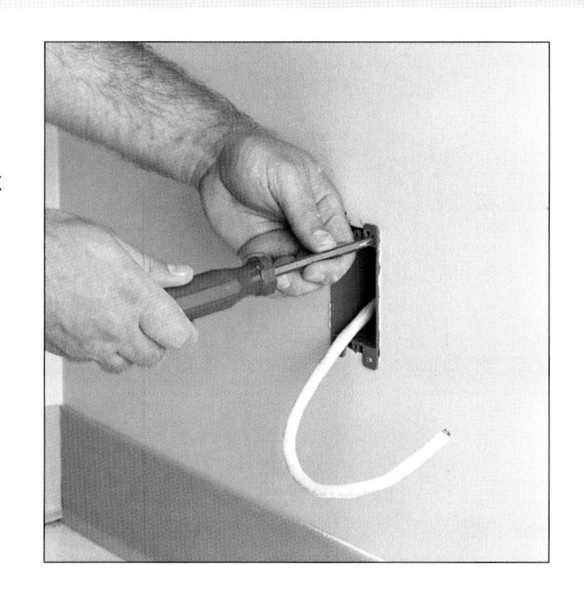

5 **Insert the box** With the cable secure, insert the box and cable gently into the hole. If it doesn't slide in easily, don't force it—you're likely to damage either the box or the wall. Use a drywall or keyhole saw to enlarge the hole slightly, and try it again. When it fits, press the box in completely and secure it by tightening the mounting screws (the box *shown* is the plastic wing type).

Common Receptacle Wiring

The most common receptacles you'll come across are side-wired receptacles. These receptacles have two pairs of screw terminals: One pair is brass or black; the other pair is silver. Only hot wires (black or red) are connected to the brass terminal; only white wires are connected to the silver (neutral) screw terminals.

Another common receptacle is the back-wired receptacle. These have openings in the rear where wires are inserted for quick installations (*see page 55*); some receptacles have both side and back wire terminals. You'll also find a link between the screw terminals that, when removed, will allow current to flow only to the outlet where the wires are attached. This if most often done for switched outlets, but it can also be used to run separate circuits to the outlet so you can operate two high-current devices off the same receptacle.

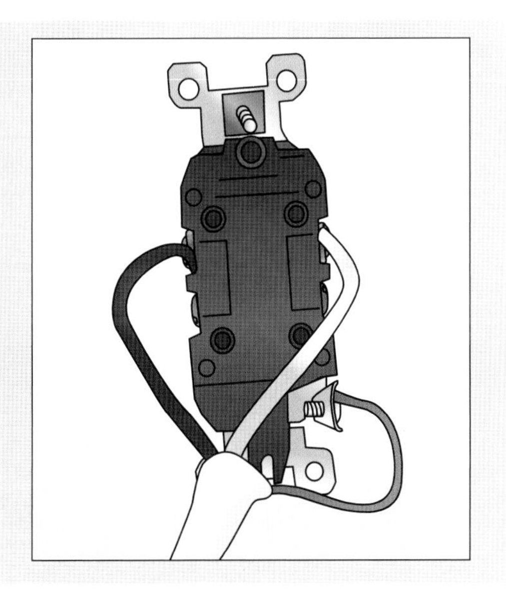

One Cable A single cable entering a box indicates end-of-run wiring—the last receptacle in the circuit. This is the receptacle you'll want to find when adding a receptacle (*see page 117*) or installing surface-mount wiring (*see page 66*). If the box is metal, the ground wire is pigtailed and connected to both the receptacle grounding screw and the box. For a plastic box, the ground wire is hooked directly to the receptacle grounding screw (*as shown*).

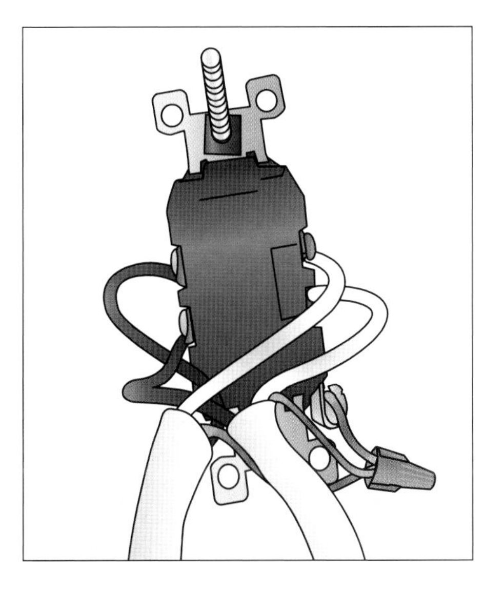

Two Cables Two cables entering a box means middle-of-run wiring—the power continues on to another circuit. The wiring is identical to a single-cable receptacle except that all the screw terminals are used and there are four wires to pigtail together for the ground. If you need to remove an outlet or outlets in a circuit, trace the wiring back to a middle-run receptacle and disconnect the appropriate cable.

Replacing a Receptacle

Although receptacles are tough and are designed to take a lot of abuse, they do wear out over time. Constant use will eventually loosen the clips inside the receptacle that grip the prongs of a plug. Hard use like yanking a cord out at an angle and inserting plugs carelessly will also shorten their life. When it's time to replace a receptacle, turn off the power before you even remove the cover plate. Then test the receptacle with a circuit tester, multimeter, or analyzer (*see page 84*) to make sure the power is truly off. Unless you're absolutely sure of the type of receptacle you need, remove the old one and take it with you to the hardware store or electrical supply house to make sure you get an exact replacement. (Certain locations in your home require a GFCI-protected receptacle—check your local code.)

1 Remove old receptacle To remove a faulty receptacle, turn off the power and make sure it's off by checking with a circuit tester—check both outlets. Then remove the cover plate screw and cover plate. Remove the mounting screws that secure the receptacle to the box. Pull the receptacle gently out of the box, and loosen each screw terminal and remove the wires one at a time.

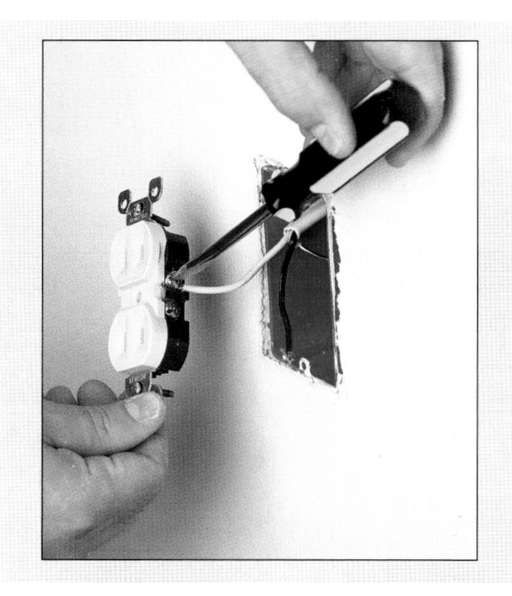

2 Install new receptacle If the existing wiring is in good shape, attach each looped wire to the appropriate screw terminal and tighten the screw. If the wires are tarnished or dirty, clean them with emery cloth or fine sandpaper before connecting them. When there's sufficient slack in the cable and you notice that the wire ends are nicked or damaged, cut them off and prepare the wire as described on *page 54 or page 55.*

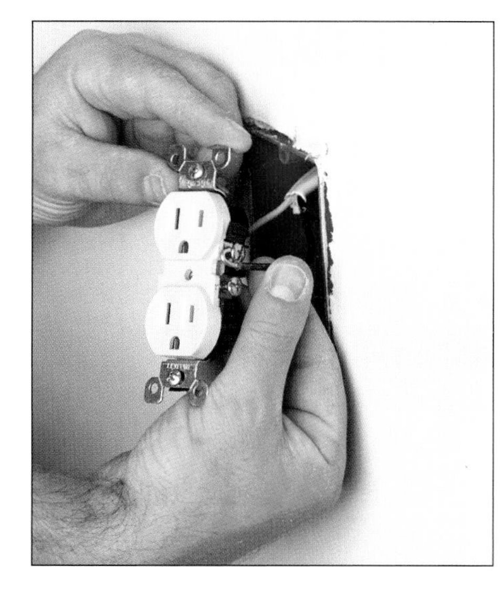

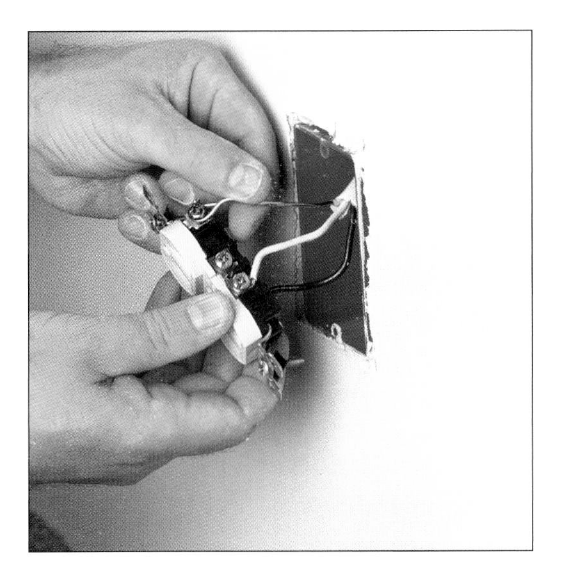

3 **Hook up the ground wire** The type of box in which you're installing the new receptacle will determine how you hook up the ground wire. For a plastic box like the one *shown,* simply hook up the bare ground wire to the grounding screw in the bottom of the receptacle. On metal boxes, run a separate ground wire to the box and pigtail this to a wire running to the receptacle's grounding screw.

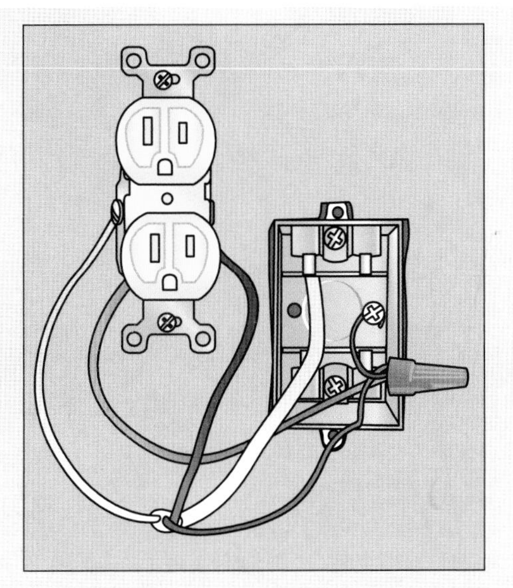

4 **End-of-run** End-of-run receptacles are the easiest to replace, as there's only one cable. Whenever you encounter one of these, take the time to make a note on your circuit map (*see pages 20–21* for more on mapping your circuits). This will be a big help on future projects when you need to add on to the run with conventional or surface-mount wiring.

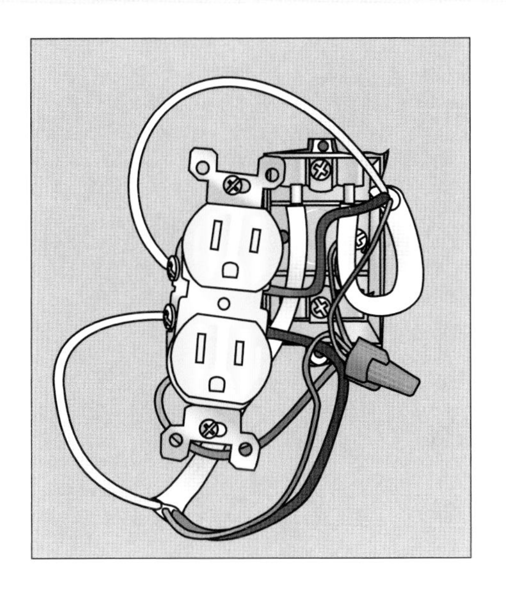

5 **Middle-of-run** A middle-of-run or mid-circuit receptacle has both an incoming and an outgoing set of circuit wires. Local codes in some areas may require that this type of circuit be joined to circuit wires with short lengths of wire pigtailed together and terminated on the receptacle. That way if there's a problem with the receptacle, power won't be interrupted to the rest of the circuit.

Adding a GFCI Receptacle

GFCI (ground-fault circuit interrupter) receptacles are safety devices designed to detect small variations in current flow between the two legs of a circuit. When an imbalance occurs, the GFCI will shut off the power to the receptacle almost instantaneously. If you're updating wiring or installing new circuits, GFCI receptacles are required by most codes in all bathrooms, kitchens, garages, crawl spaces, unfinished basements, and outdoors.

Fortunately, replacing a standard receptacle with a GFCI receptacle is simple since the connections are pigtailed together. The only challenge may be jamming the bulkier GFCI receptacle into the old box. If this is the case, consider adding a box extender to the existing box—you'll have plenty of room, and you won't run the risk of breaking a connection.

1 Remove old outlet To install a GFCI receptacle, begin by turning off power and tagging the main service panel. Then check the existing outlet with a circuit tester to make sure power is indeed off. Remove the cover plate screw and the cover plate. Unscrew the receptacle mounting screws and set them aside. Gently pull out the old receptacle, loosen each of the screw terminals, and unhook all of the wires.

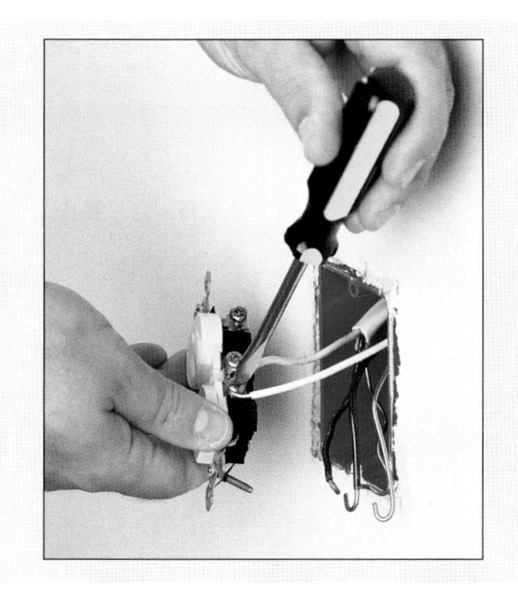

2 Pigtail the wires Cut a 6"-long piece of black, white, and green insulated wire for the pigtails. Then join all of the white wires and the white pigtail together with a wire nut. Do the same for the black wires and black pigtail and for the bare copper wires and the green pigtail. Then prepare each of the pigtail ends for the screw terminals as described on *page 54*.

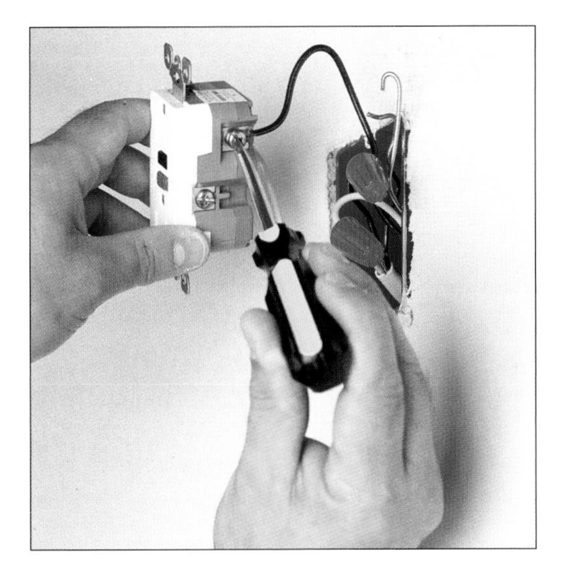

3 **Connect to GFCI** Now you can connect the pigtail ends to the appropriate screw terminals of the GFCI receptacle. If this is a single-protection receptacle, attach the pigtailed wires to the LINE screw terminals. For multiple-location protection (*see the sidebar below*), connect the incoming wires to the LINE terminal screws, and the line to be protected to the LOAD terminal screws.

4 **Connect the ground wire** If your cable provides a grounding wire or wires, connect it to the green screw terminal on the GFCI receptacle. For multiple ground wires, pigtail them together with a wire nut and connect the pigtail to the green grounding screw. Gently insert the receptacle in the box, and secure it with mounting screws. Add the cover plate and screw, and restore power. Test the receptacle as described on *page 85.*

MULTIPLE-LOCATION GFCI PROTECTION

Most GFCI receptacles can be wired to protect just themselves or all wiring, switches, and light fixtures beyond the GFCI receptacle to the end of the circuit. Note that GFCI receptacles are safest and most reliable when wired for single protection.

Multiple-location protection is susceptible to erroneous tripping when normal fluctuations occur, which can be very annoying—you'll have to reset the circuit every time this happens. If you must wire a GFCI receptacle to protect multiple locations, follow the manufacturer's directions carefully (or have a licensed electrician do the work).

Incorrectly wiring a GFCI receptacle can leave both the receptacle and the line you intend to protect without any ground-fault protection at all.

Single-Pole Switch Wiring

Single-pole switches have only two screw terminals and are used to control a single fixture or a set of fixtures from a single location. They are by far the most common type of wall switch used both in older homes and in new construction. Since they control the fixture from a single location, most single-pole switches have markings on the switch lever denoting which position is ON and which is OFF (*see page 36* for more on switch types). Basically a single-pole switch controls the hot leg of the circuit—a hot wire is attached to each terminal. When OFF, it opens or disconnects the line; when it's ON, it closes to provide a continuous path for current to flow. The color and number of wires inside the box will vary, depending on where the switch is located in the circuit; *see below.*

One Cable If only a single cable enters the box, then the switch is at the end of the run. With this type of installation (often referred to as a switch loop), one of the hot wires is black, but the other hot wire is white. If done correctly, you'll find the white wire wrapped with electrician's tape or painted black to indicate that it's hot. On metal boxes, the grounding wire should connect to the grounding screw of the box.

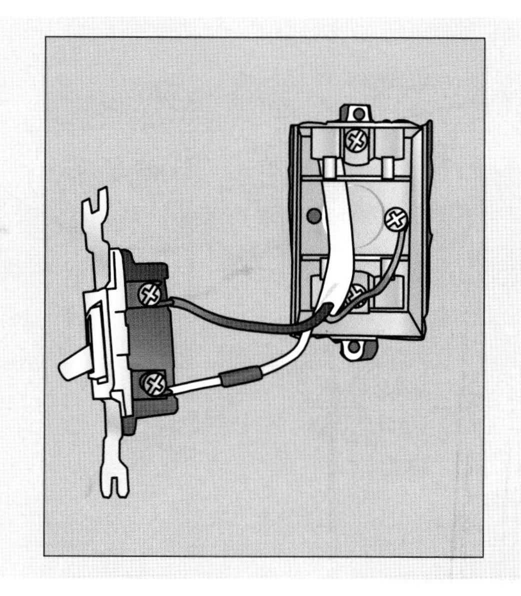

Two Cables Two cables entering the box means that the switch is in the middle of the run. Each of the incoming cables will have a black and a white insulated wire plus a ground wire. The black wires are hot and should be connected to the screw terminals on the switch. The white neutral wires are pigtailed together with a wire nut. Likewise, grounding wires are pigtailed together (and connected to the box, if it's metal).

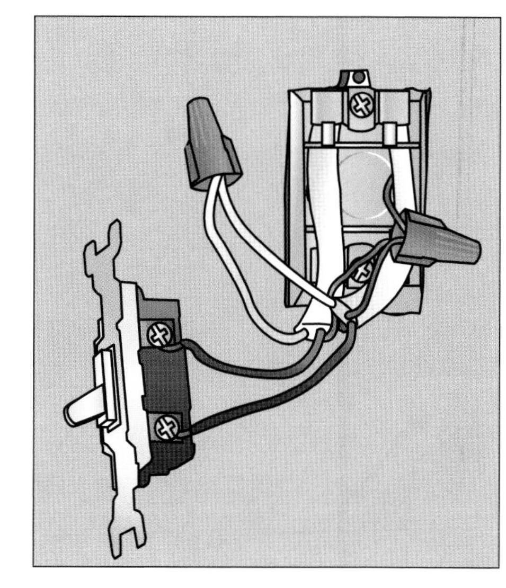

3-Way Switch Wiring

Three-way switches have three screw terminals and are always installed in pairs to control power to a fixture from two separate points (such as a light in stairway that's controlled from both the top and bottom of the stairs). Because of this, the levers on three-way switches do not have ON and OFF markings. Of the three terminals on the switch, one has a distinctive color and is marked COM for *common.* This is the terminal that the hot or black wire connects to. The other two terminals are for traveler wires that "travel" to the switches and interconnect them. (When a white wire is used as a traveler, it is marked with black tape since it carries current as well.) Three-way switches require a three-wire system—typically 12/3 with ground or 14/3 with ground, depending on the amount of current flowing through the cable (*see page 28* for more on types of cable).

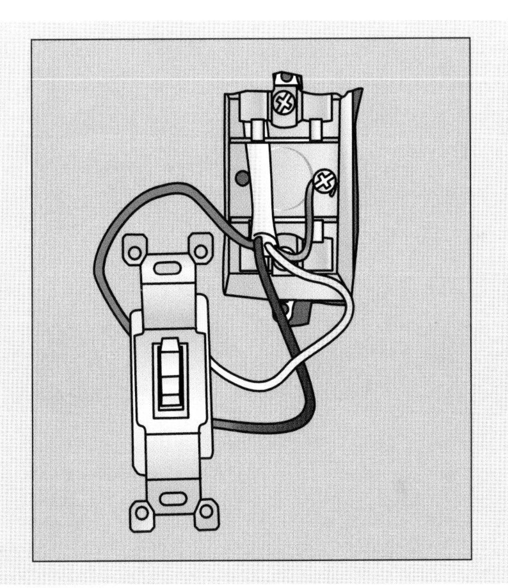

One Cable When a single cable enters the box, the three-way switch is at the end of the run. The cable should have a black, a white, and a red wire along with a ground. The black wire is connected to the common screw terminal (usually darker than the others). The white and red wires are connected to the traveler screw terminals—they're interchangeable, so either wire can be connected to either of the screw terminals.

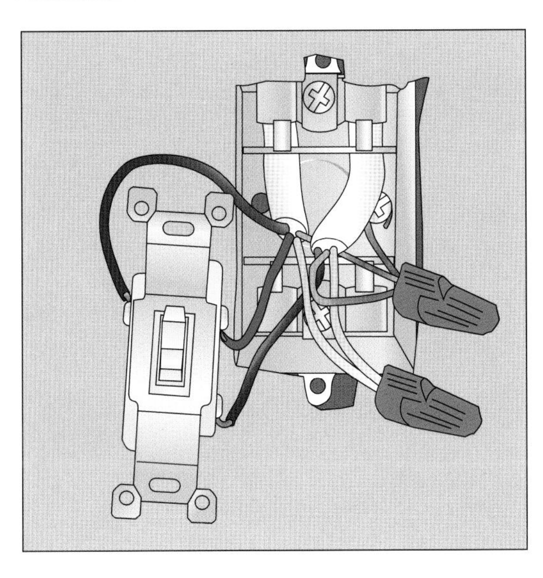

Two Cables If two cables enter the box, it's a middle-of-the-run switch. One cable will have two wires (plus ground); the other will have three wires (plus ground). The black wire of the two-wire cable connects to the common terminal. The red and black wires of the three-wire cable connect to the traveler screws. White wires are pigtailed together, and the ground wires are joined together (and connected to the box, if it's metal).

4-Way Switch Wiring

Four-way switches have four screw terminals—all travelers and no common. They are always installed between a pair of three-way switches to control a set of lights from three or more locations. Four-way switches are not common but can be found in homes with large rooms with three or more entrances or long hallways and walkways at the tops of stairs. Just like three-way switches, four-way switches don't have any ON/OFF markings.

Two pairs of color-matched wires are connected to a four-way switch in a typical installation. The screw terminals on newer-style four-way switches are color-matched in pairs to make installation easier. One of the pairs is usually copper-colored and the other pair is brass.

The idea here is to connect one set of colored wires to one pair of color-matched terminals and the other set of wires to the remaining pair. Some switch manufacturers stamp a wiring guide on the back of the switch to denote the separate lines—they're typically labeled Line 1 and Line 2.

Any number of four-way switches can be installed in a circuit, as long as they are placed in between a pair of three-way switches. Note: Take care when replacing one of these switches—it'll be worth the effort to label wires or make a simple drawing of the switch and wire positions. Attaching a set of travelers to a nonmatched set of screw terminals will result in a short circuit.

Two Sets of Wires One pair of color-matched wires is connected to the copper terminals; the other set of wires is connected to the brass screw terminals. A third pair of wires (typically black) are pigtailed together with a wire nut inside the box. The grounding wires are also joined together and, if the box is metal, are connected to the ground screw.

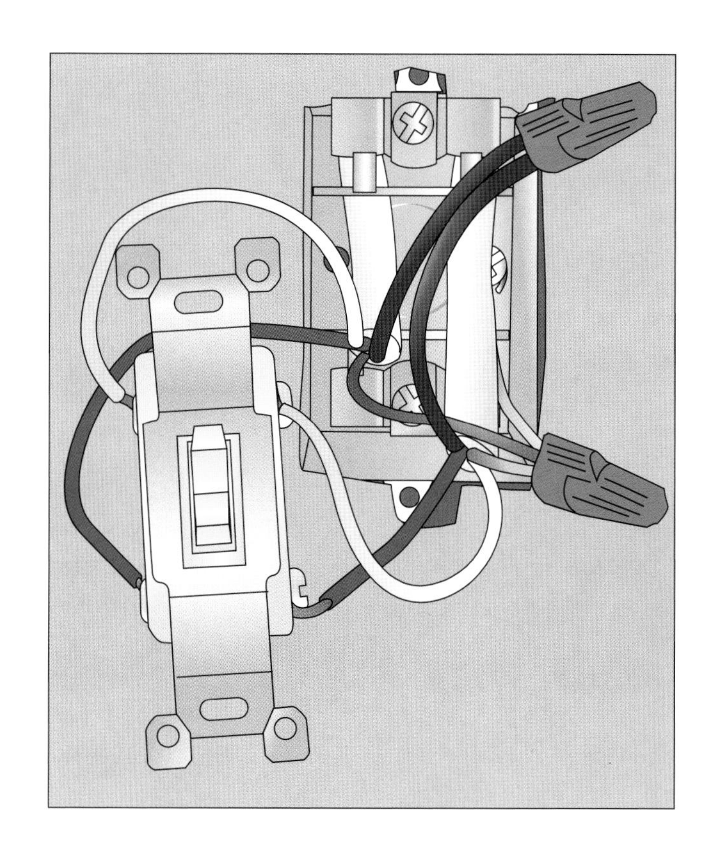

Replacing a Single-Pole Switch

1 **Remove cover plate** Switches will wear out over time and need to be replaced. To replace a single-pole switch (like the push-button style *shown*), turn off the power and tag the main panel. Then remove the cover plate screws and cover plate. If the cover plate doesn't come off easily, run the blade of a utility knife around the edge of the plate to sever any paint or wallpaper glue that may be sticking to the plate.

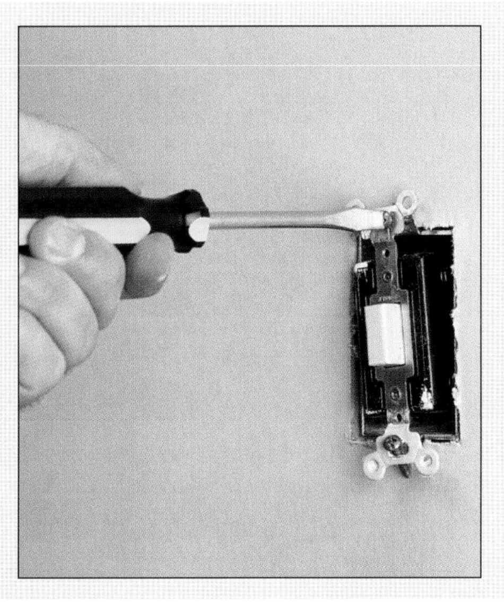

2 **Remove mounting screws** Next, remove the mounting screws that secure the switch to the electrical box. (Most mounting screws thread through small, square paper or fiberboard tabs to capture the screws in the slots of the switch.) Now pull the switch gently out of the box and test for power with a circuit tester or multimeter. Place one probe on the metal box or ground wire and touch the other probe to each screw—the lamp should not light.

3 **Disconnect and replace the switch** Loosen each of the screw terminals with a screwdriver and disconnect the wires. If the wires are tarnished or dirty, clean them with emery cloth or fine sandpaper before connecting them to the new switch. When there's sufficient slack in the cable and you notice the wire ends are nicked or damaged, cut them off and prepare the wire as described on *page 54 or page 55.*

Replacing a 3-Way Switch

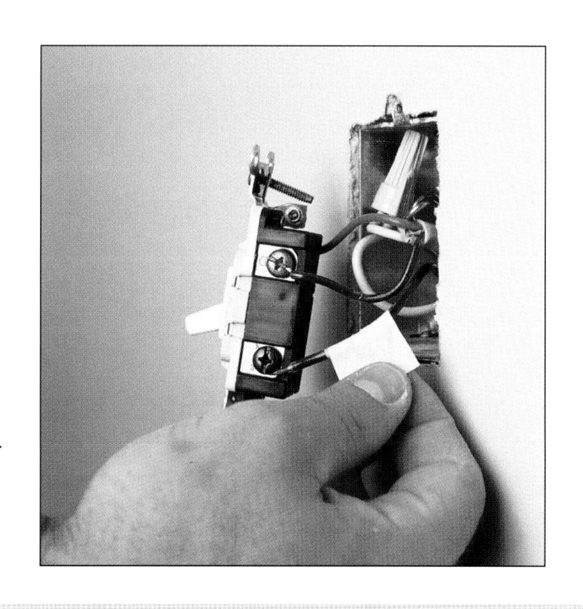

1 **Locate the common wire** Turn off power and tag the main service panel. Remove the cover plate screws and cover plate. Test for power with a circuit tester or multimeter by placing one probe on the metal box or ground wire and touching the other probe to each screw—the lamp should not light. Locate the dark common screw terminal, and mark that wire with masking tape.

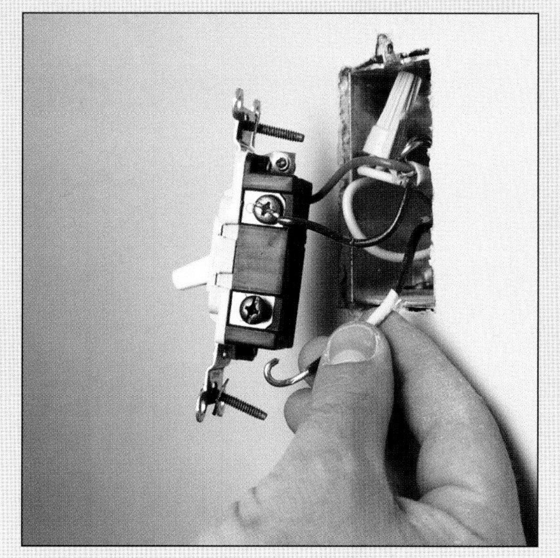

2 **Connect common wire on new switch** To make sure the new three-way switch gets wired properly, disconnect the common wire on the old switch and then immediately connect it to the common screw terminal on the new switch—it's most likely copper or black or will be labeled on the back of the switch as COMMON.

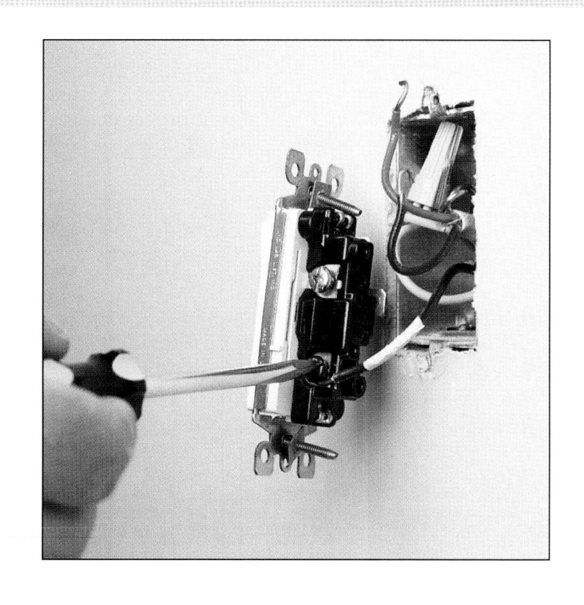

3 **Connect remaining wires** Now you can disconnect the remaining wires and hook them up to the new switch. Since these wires are interchangeable, you can attach them to either of the remaining screw terminals. When you're done, gently push the excess wire and the switch back into the box, install the mounting screws and cover plate, and restore the power. Test the switch for proper operation.

Installing a Dimmer Switch

Replacing a standard single-pole switch with a dimmer switch that can vary a light fixture's brightness is a quick and easy job. The only challenge is getting the larger-body dimmer switch to fit inside an old box. (In some cases, you'll need to add an extender to get everything to fit.) Since most dimmer switches also generate a small amount of heat, they should never be shoehorned into a small box.

Dimmer switches come in many styles: the toggle-type (*as shown here*), which resembles a standard switch; rotary-type, where rotating the dial varies brightness and the ON/OFF function (or it can be the kind that you push to turn ON or OFF); and the slide-action style, which often has an illuminated face. Most dimmer switches are designed to replace single-pole switches, but three-way dimmer switches are also available.

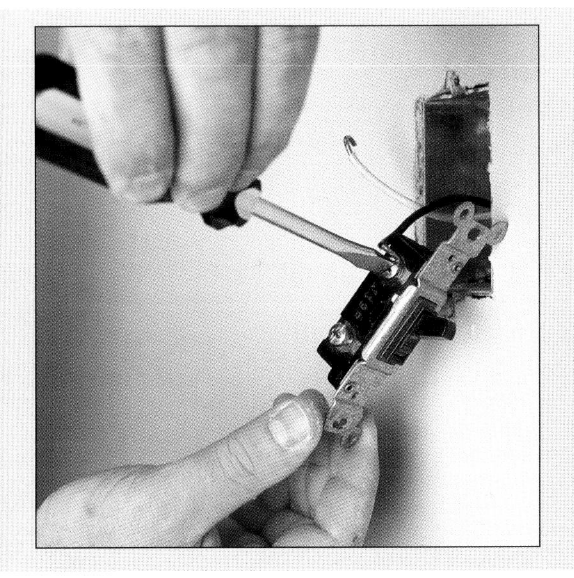

1 **Remove old switch** Turn off the power and tag the main service panel. Remove the cover plate screws and cover plate. Test for power with a neon circuit tester or multimeter by placing one probe on the metal box or ground wire and then touching the other probe to each screw—the lamp should not light. Gently pull out the switch and disconnect each of the wires.

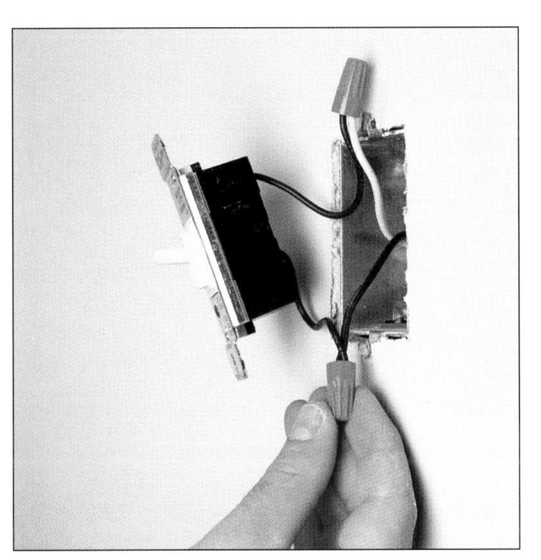

2 **Install dimmer switch** Since dimmer switches use wire leads for connections instead of screw terminals, straighten the looped ends of the wires with pliers. Connect the dimmer switch wires to the circuit wires with wire nuts (they're usually provided with the switch). On a single-pole dimmer switch, these wires are interchangeable, so you can connect them to either wire. Then gently push the switch into the box, install the mounting screws and cover plate, and restore the power.

Chapter 6

Lighting

Light at the flip of a switch is something we all take for granted. It's not until something fails that we give it a second thought. In most cases, it's simply a bulb that needs to be replaced. But if you think about how many times most fixtures are switched on and off over the course of a year, it's amazing that they perform as well as they do. Besides the constant on/off use, all light fixtures generate heat—and heat will eventually cause problems: Wires will stiffen and crack, lamp sockets will split, interior components such as starters and ballasts (for fluorescent fixtures) will degrade and stop functioning, even the metal bases and cover plates can twist and warp.

Fortunately, replacing most fixtures is simple. All it takes is an hour or two and the right replacement parts. Since lighting fixtures are heat-sensitive (some will even cut off power if they get too hot), it's imperative for the safety of your home and family that you use only the correct parts. Don't settle for something that looks like it's the right part; bring the old part with you to the hardware store or home center and ask for help, or consult a lighting specialist.

In this chapter I'll start by showing you how to replace a standard ceiling fixture (*pages 105–106*)—whether the old fixture's just not working anymore or you simply want a new look for the room. Then I'll go over replacing recessed lights (*pages 107–108*); these hidden fixtures require a bit more finesse to change but are still relatively easy. After that, there are step-by-step instructions on replacing fluorescent fixtures (*pages 109–111*) and repairing them by replacing the most common failure part, the ballast (*pages 112–113*). Finally, we'll take on replacing a common accent lighting fixture: track lighting (*pages 114–115*).

Although this chapter concentrates on replacing these fixtures, most of the instructions relate to installing them as well. The only requirement is that you have power at the location. You can either use an existing fixture or run power to the location yourself (*see pages 120–123* for more on running cable for new fixtures).

Replacing an Incandescent Fixture

1 Remove old fixture Start by turning off the power to the fixture and tagging the service panel. Then remove the globe or diffuser and the lightbulb(s). Next unscrew the retaining nut that holds the decorative cover plate onto the electrical box. If the old fixture doesn't come off easily, run the blade of a utility knife or putty knife around the edges of the cover plate to free it from paint or wallpaper paste.

2 Disconnect wiring Before disconnecting the wires, test to make sure there's no power. If the fixture is heavy, consider making a simple hook out of an old coat hanger to hang the fixture from the box while you work. Once you're sure the power is off, unscrew the wire nuts, separate the wires, and set the old fixture aside.

3 Attach the mounting plate Prior to 1959, incandescent light fixtures were often mounted directly to an electrical box. Code now requires that the fixture be mounted to a flat metal bar called a mounting strap that is secured to the box. Most new fixtures include a mounting strap (or you can buy a "universal" mounting strap at your local hardware store). Fasten the strap to the box with the screws provided. Note that here I've added a large decorative ring to compensate for the variation in base size between the old and new fixture.

4 **Install the new fixture** Before installing the new fixture, inspect the wires coming out of the box. If the insulation is cracked, wrap electrician's tape around it. If the ends are nicked, cut the ends off and re-strip. Attach the new fixture wires to the circuit wires with wire nuts that are supplied with the new fixture. (Most fixtures come with fiberglass insulation in the canopy to prevent heat from the bulb from deteriorating the wires' insulation.)

5 **Add bulbs and diffuser** Finally, screw in the appropriate bulbs and attach the diffuser. Make sure the bulbs you are using have a wattage rating less than or equal to the maximum allowable wattage rating for the fixture. The diffuser is typically held in place with a decorative cap or retaining nut. Tighten this friction-tight and no more—overtightening can crack the diffuser.

Ceiling Box Support Options

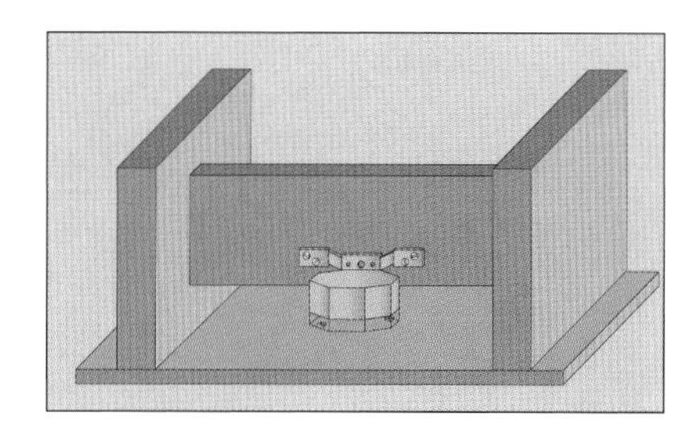

Box with Metal Brackets This type of box attaches to a 2×6 cleat set back so the box ends up flush with the ceiling.

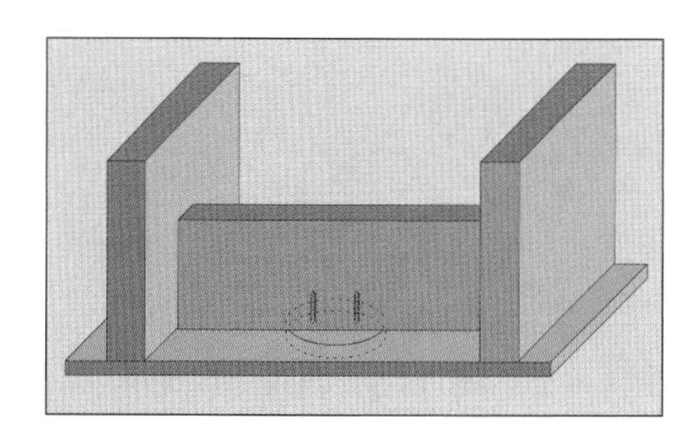

Pancake Box This is inset into the ceiling and screwed directly to a cleat fastened flush with the ceiling.

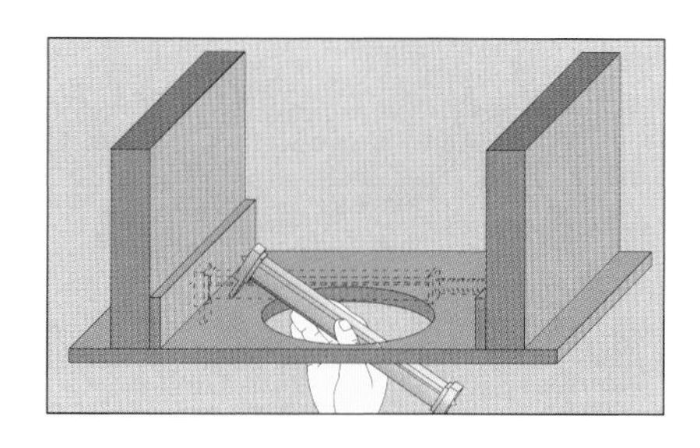

Adjustable Bracket Box An adjustable bracket is fastened to the ceiling joists. Then the box is attached to the bracket.

Replacing a Recessed Light

Since recessed light fixtures are set up into the ceiling and there's little or no air circulation, they often encounter problems because of the heat they generate. Unlike their more reliable surface-mounted cousins, recessed light fixtures will regularly fail over time and need to be replaced. The biggest challenge to replacing one of these lights is the lack of "knuckle" room inside the fixture itself. If you've got access to the fixture from above, this is a moot point—you can simply lift the fixture out for replacement.

The recessed light *shown here* attaches to a pair of sliding brackets that are fastened to the ceiling joists. Although a separate electrical box for the fixture is not needed, there's usually one nearby to serve as a transition between the power cable (typically non-metallic cable) coming into the box and the cable leading to the fixture (usually armor-clad cable).

1 Remove trim, bulb, and diffuser Begin work by turning off power to the fixture and tagging the main panel. Then unscrew the bulb and remove the trim. This conical-shaped piece made of plastic or metal is attached to the canister via a pair of spring clips. Use a pair of needle-nose pliers to reach up into the canister and lift the ends of the springs out of the slots they ride in. Pull out the trim and set it aside.

2 Loosen screws or clips The canister of a recessed light attaches to the ceiling by way of a mounting frame that is fastened either to the ceiling itself or to the ceiling joists with sliding brackets. To remove the canister, you'll need to detach it from the mounting frame—it's most often held in place with a set of three or four hex-head screws. Since space is tight, it's easiest to remove these with a socket wrench.

3 **Lift out the fixture** Once you've removed the screws, be careful because the fixture will be loose and could possibly slide right out of the ceiling. Work the canister slowly out of the mounting frame. As you get near the top of the canister where the cover plate and electrical cable attach, you'll need to tilt the canister to one side so it can slide out of the frame. Pull the canister out until you have access to the cover plate.

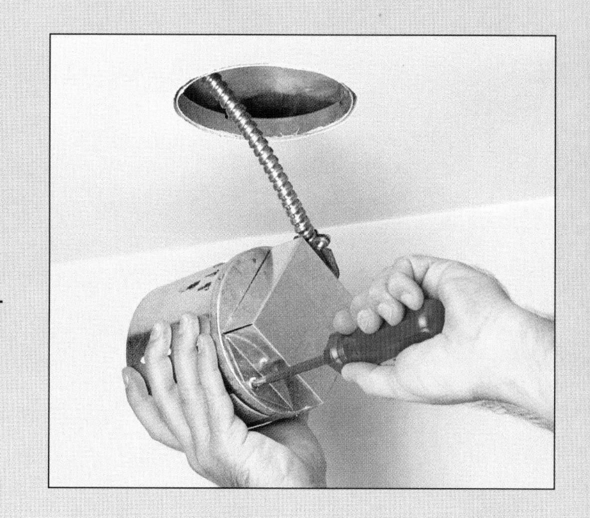

4 **Remove the cover plate** The cover plate on top of the canister provides access to the fixture wiring. It's held in place with one or two screws and is easily removed. After you've removed the cover plate, test the wires for power with a circuit tester before disconnecting them. If your canister has a permanently attached plate or cap, you'll need to disconnect the wiring back at the transition box and replace the entire unit.

5 **Install the new fixture** Take the old fixture to a lighting supplier to purchase an exact replacement. This way you can use the old mounting frame so the new installation will be a snap. Join the fixture wires to the old wiring with wire nuts: black to black, white to white, and green to ground. Replace the cover plate, feed the canister into the ceiling, and attach it to the mounting frame. Add the trim and bulb and restore power.

Replacing a Fluorescent Fixture

Fluorescent light fixtures have a number of advantages over incandescent fixtures: They run a lot cooler while providing more light; they consume less electricity; and the fluorescent tubes typically last up to three years. But that's not to say fluorescent lights don't have their problems. One of the internal parts, the ballast, has a limited life span; *see page 112.* In some cases, it may be cheaper to replace the fixture itself instead of the ballast. The procedure for replacing almost any fluorescent fixture is the same. The only difference has to do with how the fixture attaches to the ceiling. The simplest arrangement is when the fixture hangs from chains. Next is a fixture that screws directly to ceiling joists (*as shown here*). The most complex is when the fixture is recessed into the ceiling—there's less elbow room. Safety note: The metal edges of fluorescent fixtures are often sharp. Wear gloves whenever you handle them.

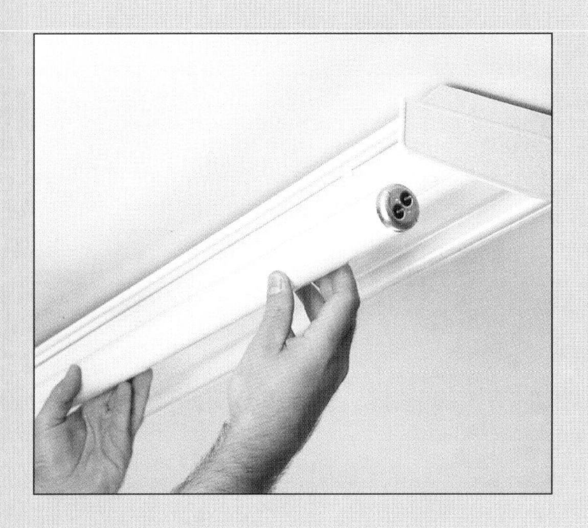

1 Remove diffuser and bulbs Turn off power to the fixture and tag the main service panel. Carefully lift off and pull out the plastic diffuser—it hooks onto the metal edges of the fixture. Be careful, as the plastic can be brittle and breaks easily. Set it aside in a safe place and then remove the light tubes. A one-quarter turn in either direction will usually release them. Carefully slide each tube out of the socket, taking care not to damage the pins on the ends of the tube.

OLD FIXTURES: STARTERS

Older fluorescent fixtures usually have a device called a starter located near one of the sockets. The starter helps to energize the gases in the tube. Starters are typically around ½" in diameter and have a pair of contacts on the bottom. With the power off, push the starter in while twisting counterclockwise to remove it. Take it with you to the hardware store or electrical supply house, and buy an exact replacement. It's a good idea to replace the light tubes whenever you replace the starter.

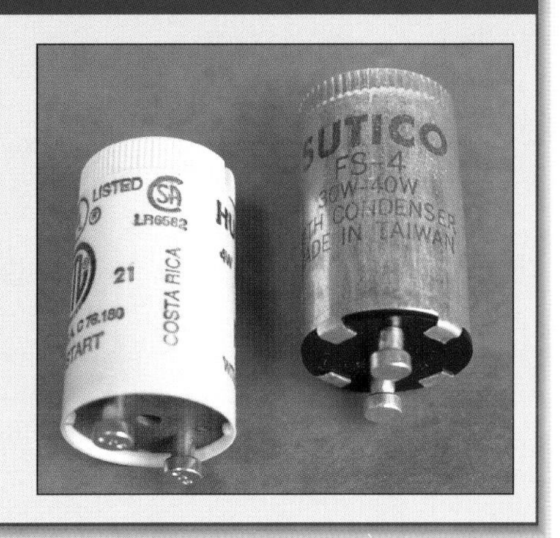

2 **Remove the cover plate** Now you can remove the cover plate to access the fixture's wiring. On some fixtures (like the one *shown here*) the cover plate is held in place with interlocking metal tabs. Just squeeze the cover plate together to disengage the tabs, and set it aside. Cover plates on other models may be attached to the fixture with screws.

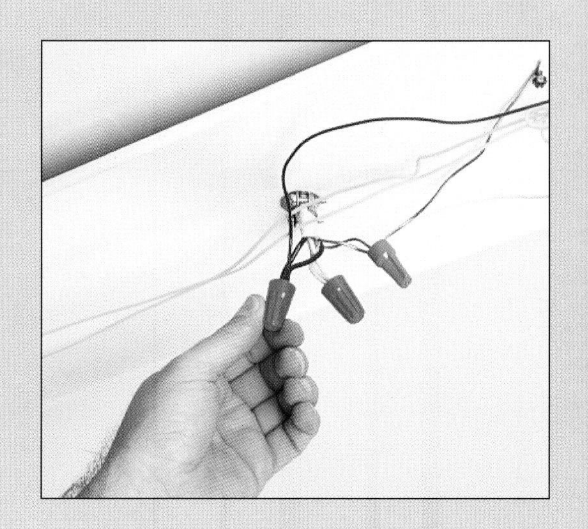

3 **Disconnect the wires** Once you've gained access to the wiring, take the time to check the wires with a circuit tester or multimeter to make sure the power is indeed off. Then unscrew each of the wire nuts and separate the fixture wires from the circuit wires. If any of the insulation is cracked, wrap electrician's tape around it or replace the wire if it's badly damaged. If the ends of the wires are nicked or tarnished, cut off the ends and strip off about 3/4" of insulation.

4 **Loosen the cable clamp** Regardless of how the fixture attaches to the ceiling (hanging on chains, screwed directly to the ceiling, or recessed into it), the cable and wires coming into the fixture should pass through a cable clamp. Before you can remove the fixture from the ceiling, you'll need to first loosen the cable clamp screws with a screwdriver.

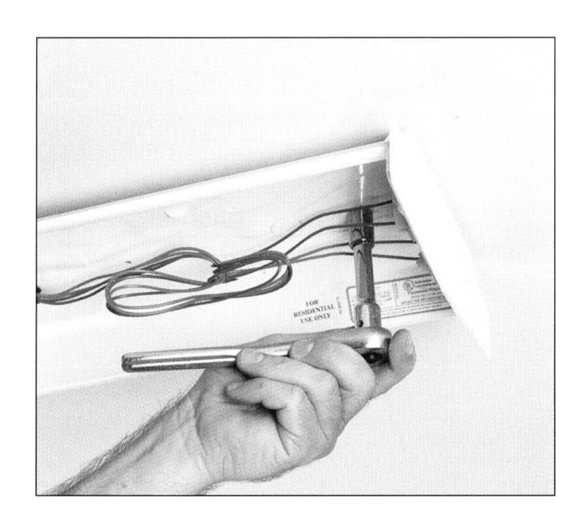

5 Detach the fixture Now you can detach the fixture from the ceiling. If it hangs on chains, simply remove the clips that hook into the cover of the fixture. If the fixture is screwed directly to the ceiling joists (*as shown here*), use a socket wrench to back out the lag screws. For a recessed fixture, remove the screws that connect the fixture to the mounting frame, and lift out the fixture.

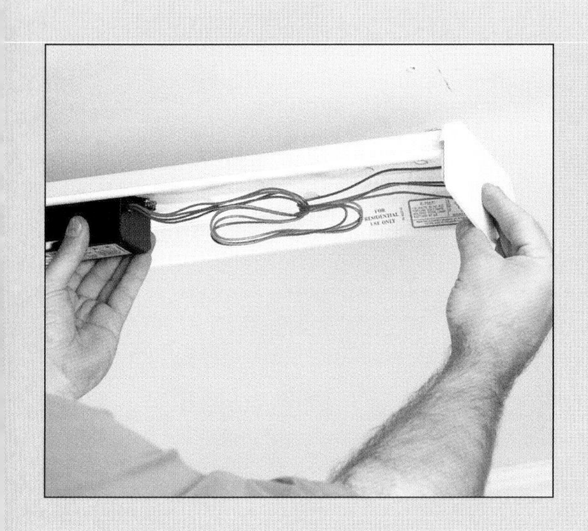

6 Remove the old fixture A helper is useful for this step because a 4'-long fluorescent light fixture is both heavy and cumbersome. Have the helper hold one side of the fixture as you remove the mounting bolts. Then carefully pull the fixture down, taking care to ease the wiring slowly through the cable clamp.

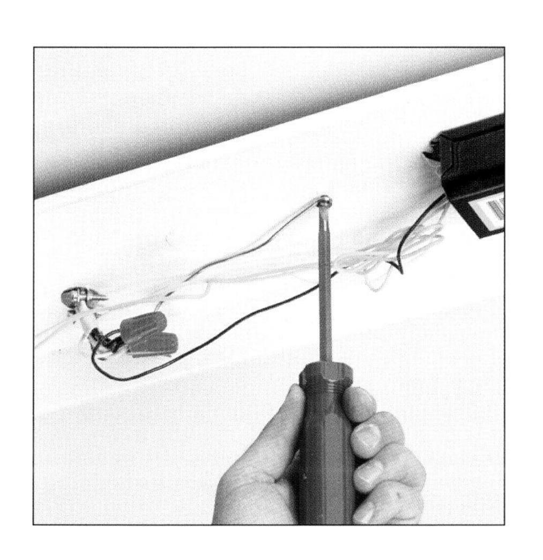

7 Install the new fixture If you've purchased an exact replacement fixture, you'll be able to use the same mounting system as for the old one; if it's a new fixture, mark and drill new mounting holes in the ceiling for bolts. Here again, a helper will make this job a lot easier and safer. Lift the new fixture into position and secure with the old hardware or use the mounting hardware provided with the new fixture. Add the cover plate, bulbs, and diffuser. Restore power and test.

Replacing a Ballast

The ballasts in fluorescent fixtures typically last around 10 years but can last significantly less, depending on the amount of use. If the fixture begins to hum and the lights flicker, first try replacing the bulbs with a set of known good bulbs. If this doesn't take care of it, odds are it's the ballast. (If you have an old fluorescent fixture, it may use a starter; *see page 109*.)

Replacing a ballast isn't difficult, but it does require the correct ballast. If you can't find what you need at the local hardware store or home center, try a lighting supplier, especially a well-established one—they often have a "graveyard" of old fixture parts in the back room. If they don't have an exact replacement, they may be able to suggest a substitute. In some cases, it may be cheaper to replace the entire fixture instead of installing a new ballast—check your local lighting supplier for a cost estimate of each before beginning work.

1 Remove the access plate Turn off power and tag the panel. You can replace a ballast while the fixture is in place, but I find that it's easier to work on if you remove it first. (*See Steps 1–7 on pages 109–111* on how to remove a fixture.) Whichever method you choose, you'll need to take off the cover plate to gain access to the ballast. The plate is often held in place with metal tabs or screws.

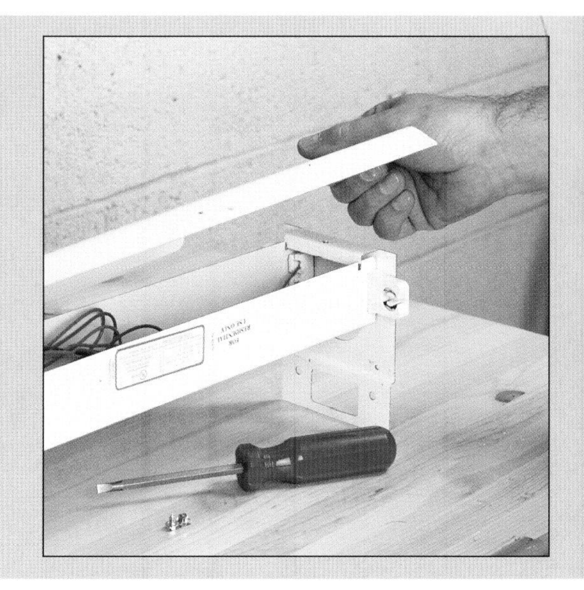

2 Remove the sockets Two pairs of color-matched wires come out the ballast and attach to the tube sockets on each end of the fixture—these need to be removed in order to replace the ballast. In most cases, you'll find it's easier to detach the wires if you first remove the sockets. Often this is as simple as lifting them out of the slots in the fixture. On other models, you may have to remove screws before you can lift them out.

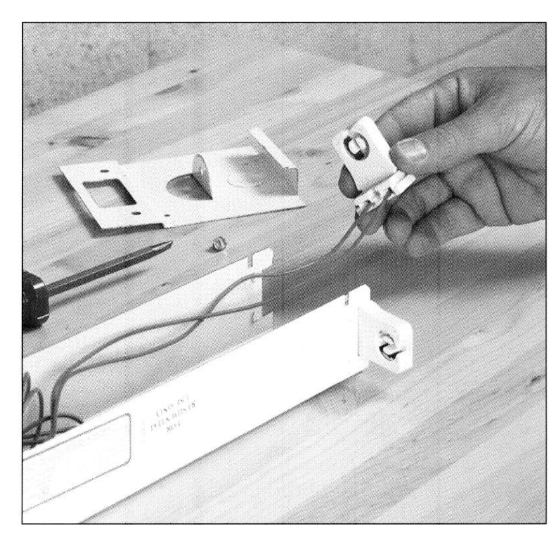

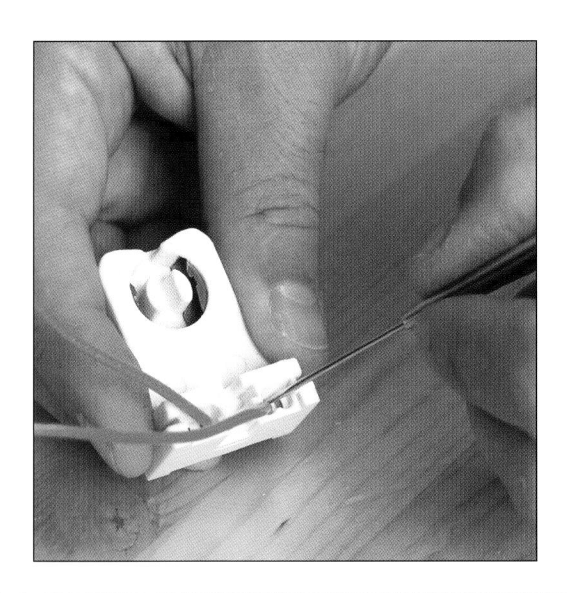

3 Disconnect wire at sockets The wires will connect to the sockets in a variety of ways: Some are screwed in place; others (like the ones *shown here*) are held in place with spring clips. To remove wires from a spring-clip socket, insert an awl or probe in between the wire and slot in the socket. This will force the spring clip away from the wire so you can pull the wire out.

4 Remove old ballast Now you can remove the old ballast. On most fluorescent fixtures, the ballast is held in place with a metal tab at one end and a screw or nut on the other. Remove the screw or nut and lift out the old ballast. Take this to your local home center or lighting specialist and get an exact replacement.

5 Install new ballast Position the new ballast in the fixture and secure it with the old mounting hardware. Attach the ballast wires to the sockets by pushing them in or using screws. If you removed the fixture, reinstall it and hook up the circuit wires to the ballast. Attach the cover plate, tubes, and diffuser. Restore power and check the fixture for proper operation.

Replacing Track Lighting

Although track lighting is a great way to provide customizable accent lighting in your home, it's the customizable part that often leads to problems. The lighting fixtures snap into the track anywhere along its length; contacts inside the base of the fixture mate with the power strip running inside the track. The problems with track lighting occur over time as the fixtures are moved around to accent different areas of the room or varying wall treatments. The contacts in the fixtures begin to fail, and the strips inside the track begin to oxidize and tarnish.

Before replacing any track lighting, try a replacement fixture. Sometimes this is all it will take. If not, replacing the track is simple. You may be able to find replacement tracks at a reputable lighting fixture store, but odds are you'll save money in the long run by replacing the entire system.

1 **Lay out the centerline** Start by turning off power to the fixture and tagging the service panel. Remove the fixtures and the track. Check the wires for power with a circuit tester before disconnecting them. Then remove the wire nuts and unscrew the old mounting plate. To install the new unit, draw a centerline parallel to the wall so it bisects the box. If the old line for the original track is still visible, proceed to Step 2.

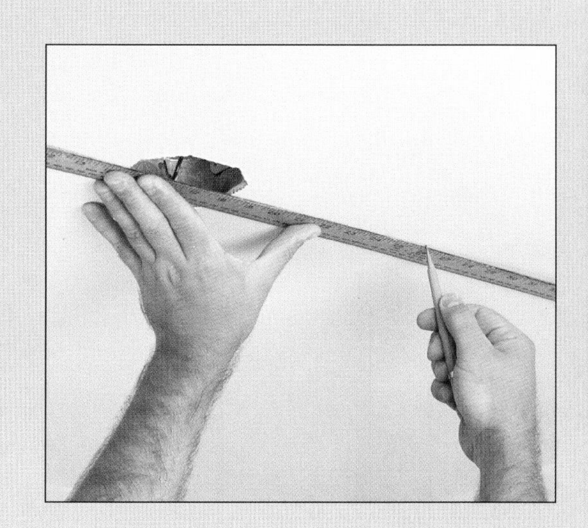

2 **Connect the wiring** Connect the wiring on the new connector housing to the circuit wiring according to the manufacturer's directions: black to black, white to white, and green to ground. Use wire nuts for quick connections.

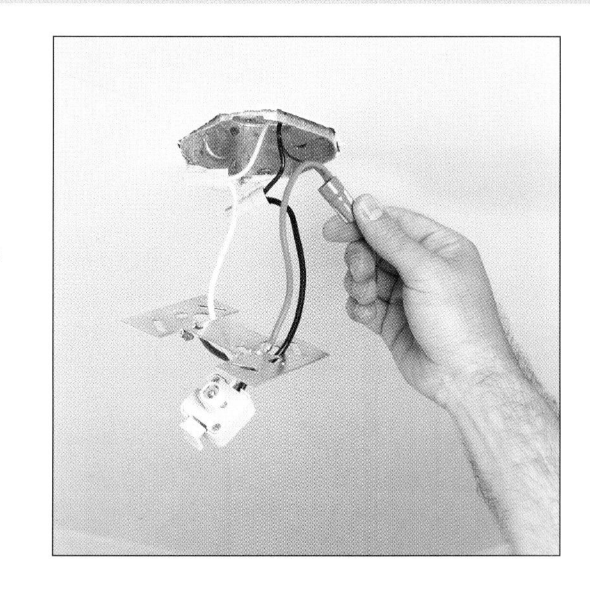

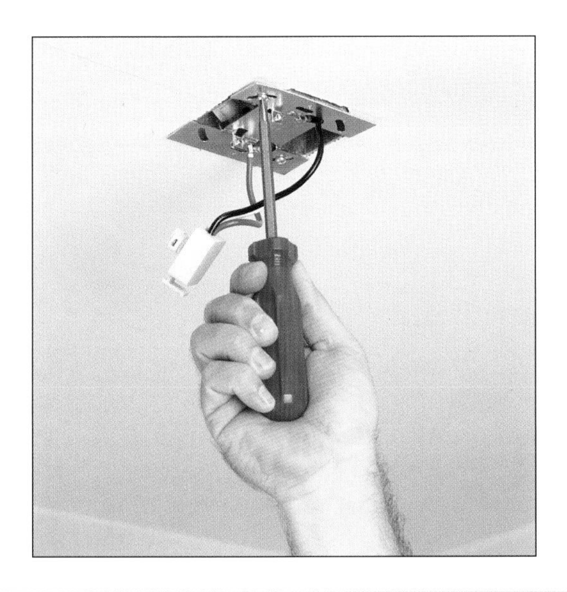

3 Attach connector to box Carefully tuck all the wires up into the electrical box and screw the connector housing to the box. Here again, installation instructions will vary from one track lighting system to another, so read the manufacturer's directions carefully and follow them to the letter.

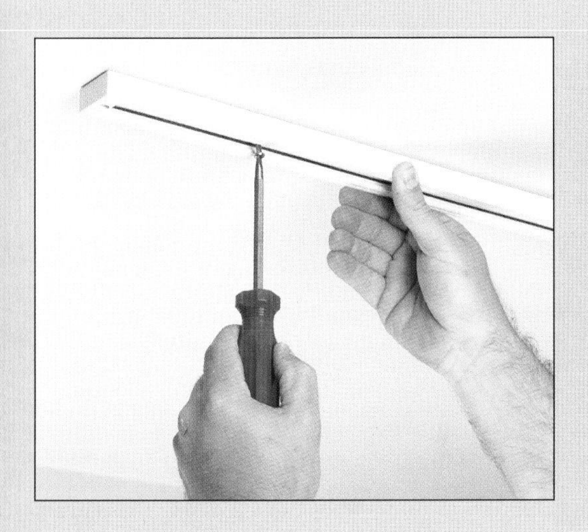

4 Install the track Align the track with the line you made in Step 1, and insert it into the connector housing. Next, make a mark through the track onto the ceiling for plastic anchors or toggle bolts. Remove the track and drill the holes for the anchors. Now hold the track in position and screw it to the ceiling. Note: Some units (like the one *shown here*) have a pair of setscrews on the connector housing to grip the track as well.

5 Attach the fixtures Before you attach the fixtures, screw on the plastic canopy that covers the connector housing. Then comes the fun part: adding the lights. Insert each new fixture into the track, slide it to the desired position, and twist it to snap it in place. Install bulbs of appropriate wattage, restore power, and test the fixture. Adjust the positions of the fixtures to get the desired lighting effect.

Chapter 7

Adding and Extending Circuits

Besides emergency repairs, adding and extending circuits is one of the more common jobs that an electrician gets called for. In situations where the job is large or complex, this is a good idea. But if all you want to do is add another receptacle to a room (and who doesn't?), or possibly need to install a new ceiling fixture, you can do the job yourself.

Before tackling an extension job, it's important that you're familiar with a couple of things. First, you must follow the electrical code and have all your planned work approved by a local building inspector. Second, you need to have a working knowledge of basic carpentry—in particular, framing. Check your local library for a basic framing book that's well illustrated and shows typical house framing. It's critical that you understand how your house is put together before you start drilling holes in it.

How you extend a circuit will depend on numerous things. First, how far the intended extension is from the main panel or the electrical box that you plan on tapping into. Second, the ability of the circuit to handle the additional load. Third, what kind of access you have under the floor of both the room you're extending to and where you're coming from. And fourth, the type of base molding in the room (I realize it sounds odd, but the size of the molding will often determine how you route the new cable).

In this chapter, I'll start with how to add a new receptacle for rooms with small base moldings (*pages 117–119*). For rooms with large base moldings, there's a trick you can do: Instead of routing the cable through the floor and back up into the new location, you can remove the base molding and "sneak" the cable behind it; *see page 120* for more on this. Then I'll show you how to run cable from a wall to a ceiling for that new ceiling fixture you bought (*page 121*). After that there's installing non-metallic (NM) cable in exposed framing—whether it's new construction or just an unfinished basement (*pages 122–123*). And finally, there's installing conduit on masonry walls: in basements, in garages—wherever the wiring can't be left exposed (*pages 124–125*).

Adding a New Receptacle

I don't know about you, but I've never lived in a house that couldn't use more receptacles. Extension cords may seem like the answer, but it's easy to overload them and run the risk of a fire. A better solution is to add receptacles so that they look like they've always been there. It's not that difficult to extend the nearest cable and recess a box into the wall. (If this makes your nervous, *see page 66* about surface-mount wiring.)

There are a couple of things to do before you can extend a circuit. First, locate an existing box that you feel is a good candidate for extension. End-of-run receptacles (*page 95*) are what you're looking for, as long as they're not controlled by a switch. Second, check with an electrician to make sure the circuit can handle the additional load. And finally, the electrical box must be large enough to accept the extra wiring (*see pages 30–32*).

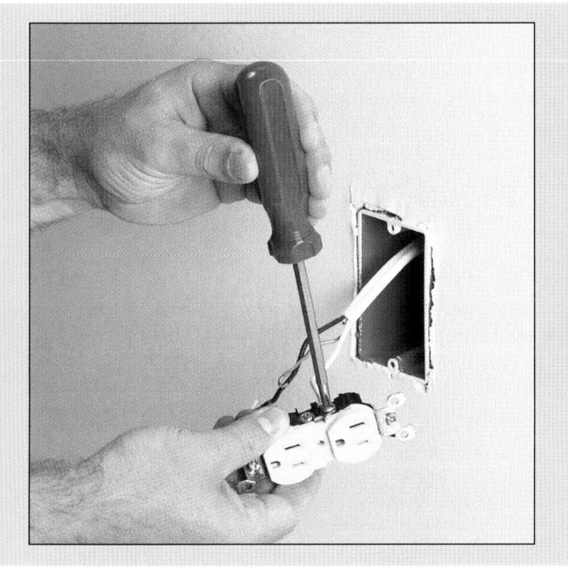

1 Access the existing box Once you've identified the receptacle you're going to extend, turn off the power and tag the main panel. Then remove the cover plate and the receptacle mounting screws. Gently pull the receptacle, and use a circuit tester or multimeter to test it to make sure power is off. Loosen the screw terminals, disconnect the wires, and set the receptacle aside.

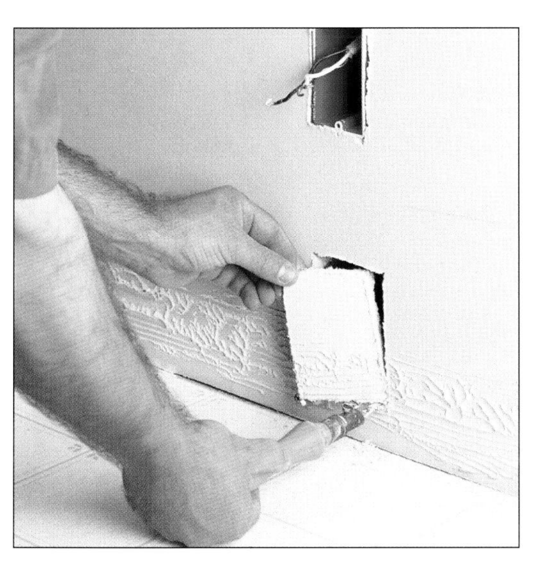

2 Cut access holes First remove any base molding. Then cut holes near the base of the new and existing receptacles to gain access to the soleplate, which you'll drill through in Step 3. The holes allow you to run the cable down through the floor and up into the new receptacle. Cut the holes with a drywall saw, roughly 3"×5" and as close to the soleplate as possible. Save the cutout piece for patching later (*see page 119*).

3 Drill holes in the soleplate Before you drill any holes in the soleplate, check the area under the floor where you'll be drilling to make sure there are no obstructions, plumbing, or gas or electrical lines near it. To drill the holes, I've found that an extension bit equipped with a $\frac{5}{8}$" spade bit works best. Center the bit on the soleplate and hold the drill at an angle as you drill to keep from damaging the wall. Do this at the new location as well.

4 Fish cable into the basement With the hole drilled in the soleplate, you can begin fishing the cable. Start by feeding the fish tape through the hole in the soleplate and down into the basement. Attach the cable to the fish tape as described on *page 64,* and pull the cable up through the soleplate. Disconnect the fish tape and feed the cable up into the existing box.

5 Fish up into the new receptacle After you've fed the cable up and into the existing box, run the cable over to the hole in the soleplate you drilled in Step 3 at the new receptacle location. (*See page 123* for options on running and securing NM cable.) Here again, you'll want to feed the fish tape through the hole in the soleplate and into the basement. Hook the cable to the fish tape, and pull it up through the soleplate.

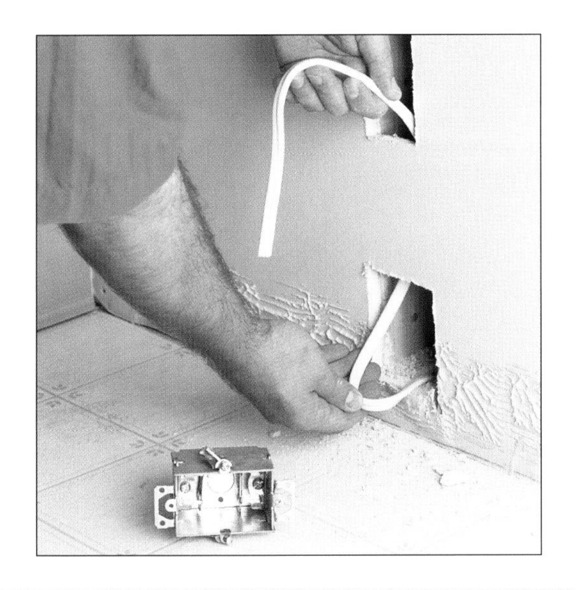

6 Feed cable to the new box Disconnect the cable from the fish tape, and feed it up through the access hole and into the hole cut for the new receptacle box. (*See page 87* for step-by-step instructions on installing electrical boxes for receptacles.) As always, make sure to leave yourself plenty of excess cable. Insert the cable into the box, secure it with the cable clamp, and install the box in the wall.

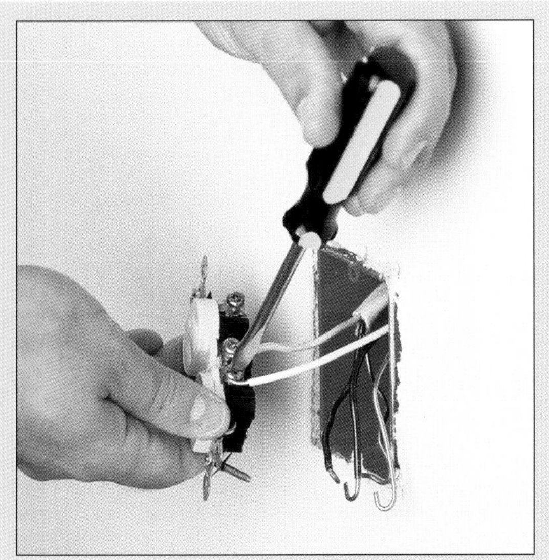

7 Connect the wiring Prepare the cable (*see page 51*) and strip the wires and get them ready for screw terminals as described on *page 54.* The old receptacle is wired as a middle-run and the new receptacle is an end-of-run. (*See page 93* for common receptacle wiring.) Press each receptacle gently into its box and secure both with mounting screws. Add the cover plates, restore power, and check for proper operation. Patch the drywall as described *below,* and reinstall the base molding.

PATCHING DRYWALL

If you saved the cutouts from the access holes you cut, you can use them as plugs for a quick-and-easy patching job. There are two quick methods of holding the plug in place. One is to use drywall clips. These clips slide onto the edge of the plug and are designed to support a patch around its perimeter. You can find these at most hardware stores and home centers.

The other method uses a scrap of wood about 1" wide and roughly 2" longer than the width of the hole. Insert the strip into the hole and fasten it with a screw on each end so it spans across the hole. Now you can screw the drywall plug to the strip.

Once the plug is in place, cover any screw heads and fill any gaps with spackling compound. When dry, sand it smooth and paint it.

Running Cable through a Wall

1 Cut holes in the wall An alternative to drilling holes in the soleplate (once you've selected a receptacle to extend—*see page 117*) is to run the cable along the wall behind the baseboards or behind cove base molding. To do this, first remove the baseboard or molding with a pry bar or a wide putty knife. Then cut a series of access holes at each stud (locate them with a stud finder), using either a drywall saw or a reciprocating saw.

2 Cut notches for cable The next step is to cut notches at the bottom of each wall stud. You only need to cut the notches deep enough to accept the cable (roughly ½" deep). The simplest way to cut the notches is to use a chisel and hammer—I suggest using an old chisel in case you hit a nail. Check each notch periodically with a short piece of cable to make sure it's deep enough.

3 Run wire Now feed the cable from the existing box to the new box by threading it through the access holes you cut in Step 1. As you pull the cable, make sure it rests in the notches that you cut in the wall studs. If the baseboard for the room is nailed in place—and even if it's not—it's a good idea to nail on metal cover plates to protect the cable from possible damage (*inset*). Patch the drywall, replace the molding, and restore power.

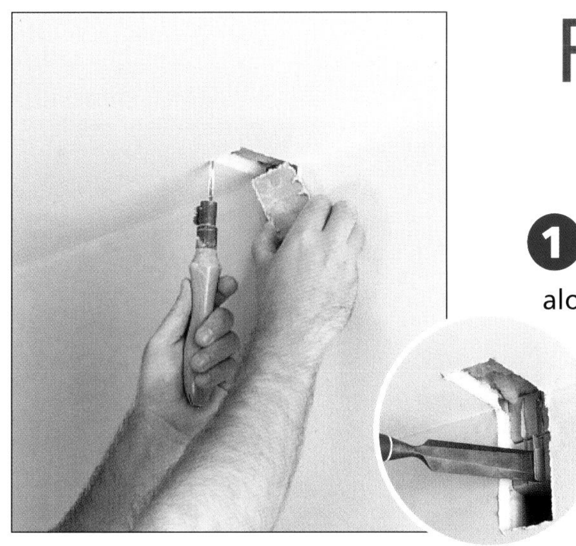

Running Cable from Wall to Ceiling

1 Cut access holes Running cable up a wall and into a ceiling for a light fixture is similar to running it along the wall (*see page 120*). The difference is the notch that you cut—use a drywall saw to cut an L-shaped hole where the wall meets the ceiling, *as shown.* Since you'll be patching this, make this notch as small as possible. Then chisel out a notch in the header for the cable (*inset*).

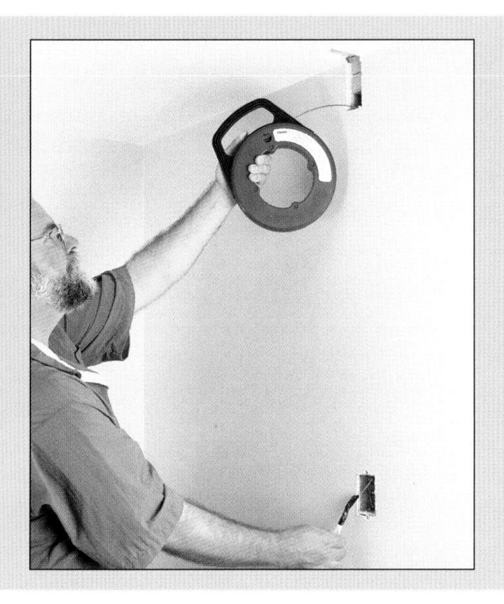

2 Fish to the ceiling Once the notch has been cut, run a fish tape down through the access hole and into the existing box. (You may need to remove a knock-out for this; *see page 88.*) Attach the cable to the fish tape (*see page 64*), and pull it up through the box and into the access hole. Pull enough cable through to reach the new fixture. (*See page 89* for information on installing boxes for light fixtures.)

3 Fish to the fixture Disconnect the fish tape from the cable and run it from the opening in the ceiling for the new fixture back to the access hole. Attach the cable to the fish tape and pull the cable to the new fixture opening. Disconnect the fish tape, insert the cable in the light fixture box, and secure the box to the ceiling. Patch the access hole (*see page 119*), install the fixture, and restore power.

Installing NM Cable in New Construction

1 Drill holes To install non-metallic (NM) cable when the framing is exposed, start by mounting the electrical boxes (*see pages 86–90* for proper heights). Then drill holes in the studs to route the cable. Use a ⅝" or ¾" spade bit, and drill the holes 1¼" back from the front edge of the stud. Try to drill the holes at roughly the same height to make pulling the cable easier.

2 Run cable Start at one end of the project (the main circuit panel or the box that you're tapping into), and thread the cable through the holes you just drilled. Continue pulling the cable until you reach the destination box. If you encounter a snag along the way, don't be tempted to pull it through with brute force—the jagged edges of a hole can easily peel the insulation off the cable.

3 Install cover plates If you discover that any of the holes you drilled in Step 1 are closer than 1¼" to the front edge of the stud, you should protect the cable against future puncture (from hanging a picture, etc.) by installing a metal cover plate. Depending on the future use of the area you're working in, you may want to install these wherever you've drilled through a stud.

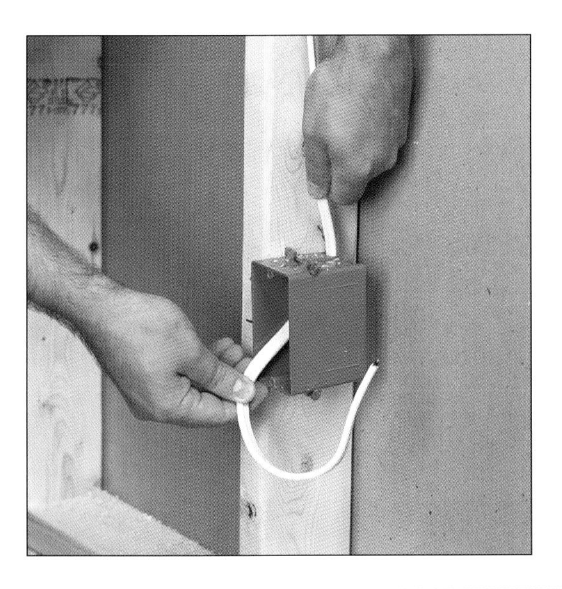

4 Run wire into the box The next step is to run the cable into the electrical boxes. Depending on the type of boxes you're using, you may or may not need to remove some knockouts (*see page 88* for more on this). Bend the cable down as it comes out of the hole in the stud, and feed it down and into the box.

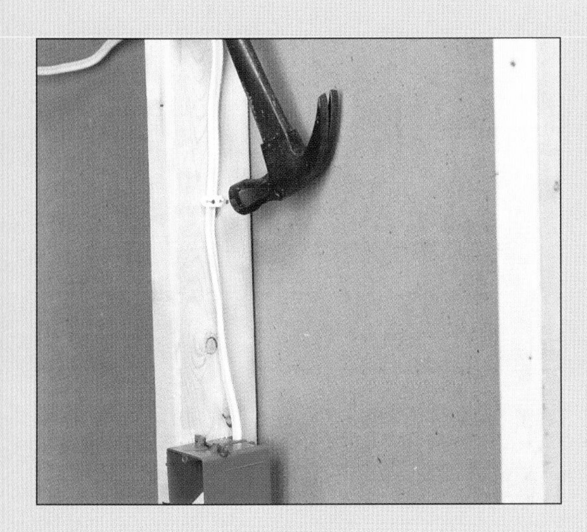

5 Staple near the box With the cable in the box, the last thing to do is to secure it. Securing a cable protects it from future damage. Code specifies that the cable be should be run in a neat manner and stapled to maintain safe distances from the finished wall. This helps prevent nails or screws from puncturing the cable. Nail the cable staples within 6" of the box to prevent this from happening.

Cable-Running Options

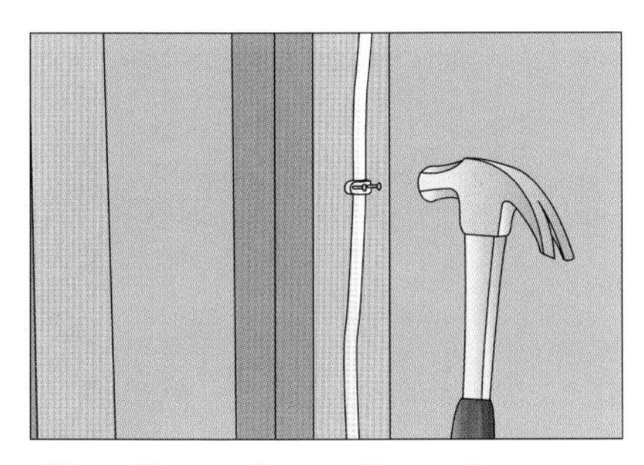

Stapling along Framing Run NM cable along a framing member, and attach it every 24" or so with cable staples.

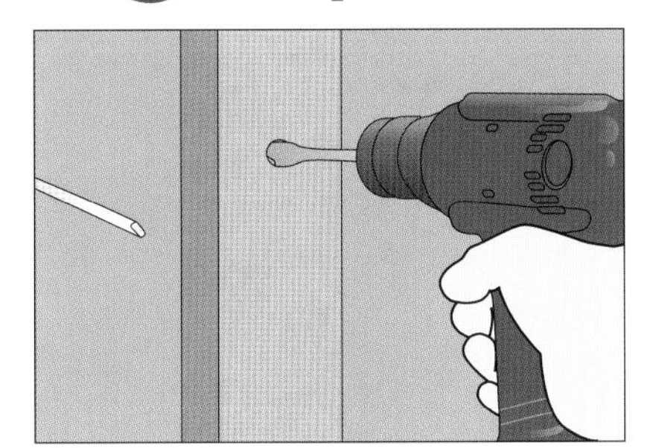

Drilling through Framing Run cable through $\frac{5}{8}$"-diameter holes drilled $1\frac{1}{4}$" in from the edge of framing.

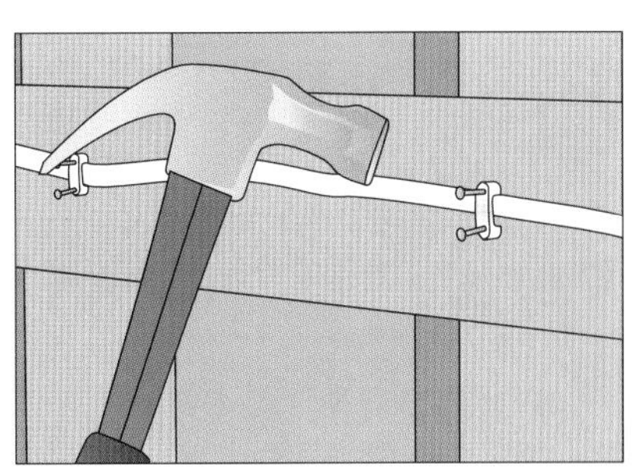

Attaching to a Cleat In areas where the walls are unfinished, staple cable to a cleat spaning the framing.

Installing Conduit on Masonry Walls

Whenever you need to run wiring in an exposed area (like in a basement or garage with masonry walls), it must be protected with conduit if it is susceptible to damage. Many homeowners shy away from working with conduit because they think it's too difficult to work with. But with the advent of some new tools and materials, it's really quite easy.

For starters, attaching boxes and conduit to masonry is easy if you use concrete fasteners (Tapcon is one brand of these). All you have to do is drill a small hole with a masonry bit sized for the fastener, and then screw the fixture into the wall—no anchors necessary. Working with conduit couldn't be easier, with the wide variety of preformed bends and the multitude of connectors available (*see pages 30–33*). A hacksaw or tubing cutter is all you need.

1 Locate and install boxes With the power off, use a torpedo level to position the boxes where you need them on the wall—check your local code for location specifications. Mark through the box onto the wall and drill holes sized to accept the fasteners you're using. Then attach each box to the wall with concrete fasteners. Don't overtighten these, since they're liable to strip—stop as soon as the fastener bottoms out on the box.

2 Add offset fittings To run the conduit flush with the walls, add offset fittings wherever conduit mates with a box. Punch out the appropriate knockouts and insert the fitting. Thread on the locknut, and tighten it with a flathead screwdriver by jamming the tip of the screwdriver against one of the locknut tabs. Push the tab in a clockwise direction.

3 Cut the conduit Measure and cut the conduit to length with a tubing cutter (*as shown*) or a hacksaw. Remove burrs from the inside edges of the pipe with the built-in or separate reamer. You can use fittings or prebent pipe sections to go around corners (*see page 30*), or you can bend the conduit yourself. (*See pages 60–63* for more on this.) Insert the conduit into the offset fittings and tighten the screws.

4 Add fittings and supports Add more fittings as necessary to extend the conduit to the next box. Insert the conduit into each fitting and tighten the screw. When all the conduit is installed, go around the room and install supports to keep the conduit from sagging (*inset*). The ones shown here require only one hole to be drilled. Then a plastic anchor is inserted in the hole to accept a pipe support that's hammered in place.

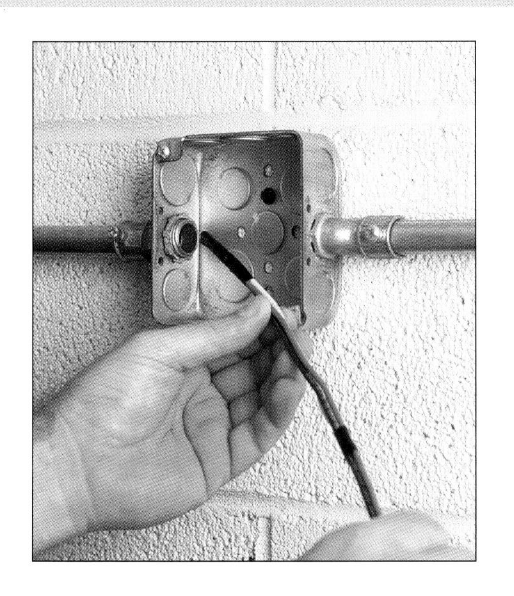

5 Fish wire through conduit Now all that's left is to fish the wire through the conduit. *See pages 64–65* for detailed instructions. Prepare the ends of the wires for screw terminals (*see page 54*), and install the receptacles (*see pages 93–95*) or switches (*see pages 98–99*). Attach cover plates, restore power, and test for proper operation.

Glossary

Alternating current (AC): the type of current found in most home electrical systems in the United States. The current continuously varies in amplitude and periodically reverses direction or "alternates" 60 times per second (i.e., at a rate of 60 hertz).
Ampere (or amp): a unit of measure for current flow that indicates the number of electrons flowing past a point in time—1 ampere of current is when 6.28×10^{18} electrons flow past a point in 1 second.
Armor-clad cable: a flexible, metal-clad cable containing two to four separate conductors; often referred to as BX, MX, or Greenfield.
Ballast: a step-up transformer that produces the starting voltage necessary to light a fluorescent light fixture.
Box: a container used to encase wire junctions or devices such as switches, receptacles, and light fixtures.
Branch circuit: a circuit that supplies a number of devices such as receptacles, fixtures, or appliances.
Bus bar: a rigid metal bar located inside the main service panel used to connect multiple conductors or devices together.
BX: *see* Armor-clad cable
Cable: two or more wires grouped together inside a protective sheathing.
Circuit: the complete path for current to flow—from a source, through devices, and back to the source.
Circuit breaker: a safety device designed to protect the wiring in your home. When an overload condition occurs, the breaker "trips" to cut off the flow of current.
Common: the identified terminal on a three-way switch—usually black or darker than the other terminals.
Conductor: any material that allows current to flow through it. Copper is an excellent conductor and is used for most wiring.
Conduit: a metal or plastic tubing that encases exposed wires to protect them.
Continuity: any uninterrupted electrical path.
Cover plate: the decorative (often plastic) plate installed over switches or receptacles to protect the wiring.
Current: the flow of electrons past a given point in a conductor in a specified time, measured in amperes.
Dimmer switch: a type of switch that allows you to vary the intensity of the light coming from a fixture.
Direct current (DC): the type of current provided by a battery; unlike alternating current, direct current does not change direction and is of constant amplitude.
Electron: the negatively charged particle in an atom; current is the flow of electrons in a conductor.
Fish tape: a flat steel spring wire with hooked ends that's used to "fish" or route cables and conductors through walls, ceilings, floors, conduit, and electrical boxes.
Four-way switch: a switch placed between three-way switches to control a single fixture from multiple locations.
Fuse: a safety device that monitors the flow of current; when too much current flows through a fuse, a thin metal strip melts—the fuse "blows"—to stop the flow of current.
Ganging: joining together two or more metal electrical boxes for greater capacity.
Ground: a conducting connection between an electrical circuit and the earth.
Ground fault: a condition that exists when the current flowing into and the current flowing out of a circuit are not the same; some current is flowing outside the circuit where it is not intended to flow.
Ground-fault circuit interrupter (GFCI): an electronic device that continuously monitors a circuit for ground faults; when one is detected, the GFCI will shut off power almost instantaneously.
Grounding conductor: usually a bare copper or green-insulated wire that provides a safe path for dangerous current to ground.
Hot wire: any conductor that brings current to a device; hot wires are covered with black or red insulation.

Insulator: a nonconductive material (usually plastic or rubber) that impedes the flow of current; conductors are encased in insulation for protection.
Knockout: a die-cut impression in a metal electrical box designed to be removed to provide an opening for cable access.
Load: a user of electricity, such as a light fixture, a microwave oven, or a television set.
Multimeter: an analog or digital measuring device capable of measuring current (in amperes), resistance (in ohms), and voltage; its test leads provide a connection to the circuit. Also called a multitester.
National Electrical Code: a body of regulations that define safe electrical procedures. Local codes often add to or modify these rules.
Neutral wire: any conductor that returns current from a device to the source; neutral wires are covered with white or light gray insulation.
Non-metallic (NM) cable: a plastic-sheathed cable that contains two to four conductors plus a bare copper ground; often referred to by the brand name Romex.
Overload: a demand for more current than a circuit can safely handle; an overload usually causes a circuit breaker to trip or a fuse to blow.
Pigtail: a short length of wire commonly used to make connections within an electrical box.
Polarized device: any device designed to ensure that current flows in the proper direction. A polarized receptacle has a short (hot) slot and a longer (neutral) slot and accepts a polarized plug, which has one short prong and one longer prong.
Power: the rate at which work is done, measured in watts; a 100-watt lightbulb consumes power at a faster rate than a 50-watt bulb.
Receptacle: a device designed to accept plug-in devices and provide them with power; also referred to as an outlet.
Resistance: the opposition to the flow of current, measured in ohms. A "load" (such as a lamp or radio) resists the flow of current—this resistance is what allows work to be done.
Screw terminal: the threaded screw found on switches and receptacles, used to make wire connections.
Service panel: a metal box that distributes power to individual circuits in the home that are protected by fuses or circuit breakers.
Short circuit: a fault that occurs when two current-carrying wires make contact; or when two conductors of different potential come together.
Single-pole switch: a switch that controls a light fixture from a single location.
Splice: an electrical connection joining together two or more wires.
Switch: a device that controls the flow of current by opening and closing the hot conductor leading to a device.
Three-way switch: a type of switch always used in pairs to control a light fixture from two different locations.
Transformer: an electrical device consisting of coils of wire designed to raise (step up) or lower (step down) the voltage coming into it.
Traveler wire: a conductor that provides a path for "hot" current between three- and four-way switches.
Underwriters Laboratory (UL): an organization that tests electrical devices for safety.
Voltage: the electrical "pressure" that causes current to flow in a conductor, measured in volts; 1 volt is the pressure required to move 1 amp through a conductor that has a resistance of 1 ohm.
Wattage: the amount of power a device needs or consumes, measured in watts; watts can be calculated by multiplying the voltage by the current.
Wire nut: a cone-shaped plastic cover used to join together and protect conductors; typically has a square-cut spring inside to grip and secure the conductors.

Index